Employment Growth from Public Support of Innovation in Small Firms

Employment Growth from Public Support of Innovation in Small Firms

Albert N. Link
John T. Scott

2012

W.E. Upjohn Institute for Employment Research
Kalamazoo, Michigan

Library of Congress Cataloging-in-Publication Data

Link, Albert N.
Employment growth from public support of innovation in small firms / Albert N. Link, John T. Scott.
p. cm.
Includes bibliographical references and index.
ISBN 978-0-88099-385-2 (pbk. : alk. paper) — ISBN 0-88099-385-5 (pbk. : alk. paper) — ISBN 978-0-88099-386-9 (hbk. : alk. paper) — ISBN 0-88099-386-3 (hbk. : alk. paper)
1. Small business—United States. 2. Research, Industrial—United States. 3. Technological innovations—Economic aspects—United States. 4. Job creation—United States. 5. Labor market—United States. I. Scott, John T., 1947– II. Title.
HD2346.U5L56 2012
331.12'5—dc23

2011049011

Cover design by Alcorn Publication Design.
Index prepared by Diane Worden.
Printed in the United States of America.
Printed on recycled paper.

For Carol and Nancy.

Contents

Figures

Tables

Acknowledgments

Our sincere thanks go out to the many individuals who were directly and indirectly involved in the research, writing, and publication of this book. We greatly appreciate the National Research Council of the National Academies for providing us with access to the Small Business Innovation Research (SBIR) database. We also appreciate the generosity of the W.E. Upjohn Institute for Employment Research in sponsoring this research project. In addition, the comments and suggestions of Randall Eberts, Kevin Hollenbeck, and Benjamin Jones, all of the Institute, and of anonymous referees have greatly improved the content and focus of this book. Of course, we are grateful for the support of our wives, Carol and Nancy, throughout this project; their encouragement has been of great value and indeed necessary for the project's success.

1
Introduction

THE PREVAILING ECONOMIC STAGE

The U.S. economy is slowly recovering from what has been the longest and arguably the most severe economic downturn since the Great Depression. The National Bureau of Economic Research (NBER) dates this recession as having lasted from December 2007 to June 2009—a period of 18 months.[1] The civilian unemployment rate rose over that period from 5 percent to 9.5 percent, and it increased to and remained at a double-digit level throughout the rest of 2009.[2]

The George W. Bush administration, with the support of the U.S. Congress, took some immediate steps to strengthen the economy. The Emergency Economic Stabilization Act of 2008 (Public Law [P.L.] 110-343), signed in October 2008, permitted the government to purchase failing bank assets in an effort to stabilize financial markets in the wake of the subprime mortgage crisis and to pay for these assets through the newly created Troubled Asset Relief Program (TARP).

The American Recovery and Reinvestment Act (ARRA) of 2009 (P.L. 111-5), also known as the Recovery Act, was signed by newly inaugurated President Barack Obama in February 2009. The premise of this fiscal stimulus was that government spending would quickly lead to business investments and subsequent consumer spending, thus mitigating the current recession.[3] Two of the stated purposes of ARRA are to "preserve and create jobs and promote economic recovery." Another stated purpose of the act is to "provide investments needed to increase economic efficiency by spurring technological advances," and throughout the act there is an emphasis on small-business investment.[4]

In September 2009, the National Economic Council (2009) released *A Strategy for American Innovation: Driving towards Sustainable Growth and Quality Jobs*. The report, put out by the Executive Office of the President, lays "the foundation for the innovation economy of the future" (p. i). This initiative "seeks to harness the inherent ingenuity of the American people and a dynamic private sector to ensure that the next [economic] expansion is more solid, broad-based, and beneficial than previous ones" (p. i). Under the heading "A Vision for Innovation, Growth, and Jobs," the report states that "innovation is essential for creating new jobs in both high-tech and traditional sectors . . . A more innovative economy is a more productive and faster growing economy, with higher returns to workers and increases in living standards . . . Innovation is the key to global competitiveness, *new and better jobs* [emphasis added], a resilient economy, and the attainment of essential national goals" (p. 4).

To accomplish this, *A Strategy for American Innovation* emphasizes that the economy must invest in the building blocks of American innovation by, among other things, restoring American leadership in fundamental research. Such building blocks will promote competitive markets that spur productive entrepreneurship, and that in turn will catalyze breakthroughs for national priorities.

Three broad and related themes are evident from the economic policies of the past several years: 1) public-sector and private-sector investments in research and development (R&D) support the innovation process; 2) innovation is closely tied to entrepreneurial activity, and new and existing small firms are more entrepreneurial than large firms; and 3) entrepreneurship and innovation are the drivers of competitiveness, new jobs, productivity growth, and overall economic well-being.

The Obama administration is not the first to embrace these themes in response to an economic crisis. In the early 1970s, and then again in the late 1970s and early 1980s, productivity growth in the U.S. industrial economy declined precipitously, thus rendering many sectors of the economy vulnerable to global competitive forces. While many explanations have been offered as to the cause of the so-called productivity slowdown, one clear policy response was to increase the level, effectiveness, and diffusion of private-sector industrial R&D and innovative activity through a series of policy initiatives.[5] These initiatives included the passage of the Bayh-Dole Act of 1980; the R&E Tax Credit of 1981;[6] the Small Business Innovation Development Act of 1982, which created the Small Business Innovation Research program (the empirical focus of this book); and the National Cooperative Research Act of 1984.[7]

AN INNOVATION-BASED ECONOMIC GROWTH STRATEGY

The three themes listed above are not without an academic foundation. Before we discuss the related literature, some initial definitions might be in order. First, invention is the creation of new knowledge, technology is the application of new knowledge, and innovation is the commercialization of the new technology—innovation puts a new technology to general use. And second, entrepreneurship involves the perception of new opportunities and the ability to act on those perceptions.[8]

The premise that the rise of technology leads to productivity growth can be traced historically to Adam Smith's *The Wealth of Nations*. A careful reading of his argument leads one to conclude that improvements in technology result from the division of labor, and a greater division of labor stimulates productivity growth. Productivity growth in turn leads to overall economic growth.

Abramovitz (1956) showed, through his pioneering analysis of aggregate economic activity in the post–Civil War economy (covering from 1869–1878 to 1944–1953), that productivity growth at the aggregate level was not due to an increase in the availability and use of resources but rather to growth in the stock of knowledge, which in turn was due to research and education. Following Abramovitz, Solow (1957) formalized the study of productivity growth, and he concluded that between 1909 and 1949 gross output per man-hour doubled, with nearly 90 percent of the increase due to technical change. This pioneering Solow study has motivated volumes of research. Nearly all of that literature substantiates the impact of investments in technology— research and development (R&D) investments in particular—on productivity growth.[9]

Regarding the role of small firms, we first focus on the scholarship of Jewkes, Sawers, and Stillerman (1958). They were among the first scholars to shed light on industrial R&D as a source of technological progress. Of the 61 important inventions they name that were made in the United States and Great Britain during the first half of the twentieth century, more than half could be attributed to entrepreneurs working on their own (i.e., as small firms) without the research resources of large firms or laboratories. More recently, in research more focused on small- versus large-firm behavior, Acs and Audretsch (1990) show that the average number of innovations per 1,000 employees made by small firms (i.e., firms with fewer than 500 employees) in the U.S. manufacturing sector in 1982 was 0.309, compared to 0.202 for large firms.[10]

In that same vein, the National Science Foundation's (NSF's) *Survey of Industrial Research and Development*, as summarized by Rausch (2010), shows that small R&D firms invest more in R&D relative to their size than do larger firms.[11] In 2007, the ratio of R&D to firm sales in small R&D firms was 8.6 percent, compared to 3.4 percent in large firms.[12]

Finally, a study by Breitzman and Hicks (2008) under the sponsorship of the Small Business Administration's (SBA's) Office of Advocacy concludes that small technology firms obtain more patents per employee than large firms. Over the 2002–2006 period, small technology firms received 26.5 patents per 100 employees, compared to 1.7 for large firms. On a patent-per-employee basis, the ratio was 1.89 for technology firms with fewer than 25 employees and 0.014 for firms with 25,000 or more employees.

At a nonstatistical level, Baumol, Litan, and Schramm (2007) emphasize the creative individual—the entrepreneur, which to many is synonymous with a newly created firm—as the originating force behind innovation. They suggest that "if the United States wishes to continue enjoying rapid growth, it must find a way both to launch and promote the growth of innovative entrepreneurial enterprises and to ensure that the successful entrepreneurs who grow their [small] businesses into large firms continue to innovate. We believe that this requires policies that encourage what we call 'productive entrepreneurship'" (p. 2).

Introducing a more nuanced and balanced view, Baumol (1990, p. 3) has also argued that not all entrepreneurial efforts are productive. He writes the following:

> Entrepreneurs are always with us and always play *some* substantial role. But there are a variety of roles among which the entrepreneur's efforts can be reallocated, and some of those roles do not follow the constructive and innovative script that is conventionally attributed to that person. Indeed, at times the entrepreneur may even lead a parasitical existence that is actually damaging to the economy. How the entrepreneur acts at a given time and place depends heavily on the rules of the game—the reward structure in

> the economy—that happen to prevail. Thus . . . it is the set of rules and not the supply of entrepreneurs or the nature of their objectives that undergoes significant changes from one period to another and helps to dictate the ultimate effect on the economy via the allocation of entrepreneurial resources.

Baumol, Litan, and Schramm (2007) argue for policies to encourage productive entrepreneurship by suggesting a number of policy initiatives that should be considered, including everything from tax policies to encourage risk taking, to reform of bankruptcy laws to promote the formation of new businesses, to revision of the patent system. Such policy actions will "adjust the rules of the game to induce a more felicitous allocation of entrepreneurial resources" (Baumol 1990, p. 4).

It should be noted that not all are in agreement with Baumol and his colleagues. Friedman (2007, p. 4), for example, is critical of this point of view, and he has argued that "implementation of changing technologies shifts the balance of skills demanded in the market for heterogeneous labor, and if the change is sufficiently rapid the mix of skills that the labor force supplies cannot keep pace. [And], advancing technology is also rendering an ever wider array of goods and services conveniently tradable across international boundaries, thereby exposing American workers to the actual or incipient threat of competition from workers willing to accept extremely low wages compared to American standards."

Academic and empirical evidence aside, there is also a well-developed theoretical foundation for government support of innovation, as is discussed next.

GOVERNMENT'S ROLE IN INNOVATION

One theoretical basis for government's role in market activity is the concept of market failure. Market failure is typically attributed to market power, imperfect information, externalities, and public goods. The explicit application of market failure to justify government's role in innovation—in R&D activity in particular—is a relatively recent phenomenon within public policy.

Public interest in innovation can be traced back to President Washington's address to Congress in 1790: "The advancement of agriculture, commerce, and manufactures, by all proper means, will not, I trust, need recommendation; but I cannot forbear intimating to you the expediency of giving effectual encouragement, as well to the introduction of new and useful inventions from abroad, as to the exertions of skill and genius in producing them at home . . . " (PBS 2002).[13]

Still, many point to the first President Bush's Office of Science and Technology Policy, which issued a document titled *U.S. Technology Policy* in 1990 as the United States' first formal domestic technology policy statement. While it was an important initial policy effort, it failed to articulate a foundation for government's role in technology and innovation. Rather, it implicitly assumed that government had a role, and then set forth the general statement, "The goal of U.S. technology policy is to make the best use of technology in achieving the national goals of improved quality of life for all Americans, continued economic growth, and national security" (Executive Office of the President 1990, p. 2).

President Clinton took a major step forward from the 1990 policy statement in his 1994 *Economic Report of the President* by articulating first principles about why government should be involved in the technological process (Council of Economic Advisers 1994, p. 191): "The goal of technology policy is not to substitute the government's judgment for that of private industry in deciding which potential 'winners' to back. Rather, the point is to correct market failure."[14]

Subsequent Executive Office policy statements have echoed this theme; *Science in the National Interest* (Clinton and Gore 1994) and *Science and Technology: Shaping the Twenty-First Century* (Executive Office of the President 1997) are among the examples. President Clinton's 2000 *Economic Report of the President* elaborated upon the concept of market failure as part of U.S. technology policy: "Rather than support technologies that have clear and immediate commercial potential (which would likely be developed by the private sector without government support), government should seek out new technologies that will create benefits with large spillovers to society at large" (Council of Economic Advisers 2000, p. 99).

Related to this, Martin and Scott (2000, p. 438) observed that "limited appropriability, financial market failure, external benefits to the production of knowledge, and other factors suggest that strict reliance on a market system will result in underinvestment in innovation, relative to the socially desirable level. This creates a *prima facie* case in favor of public intervention to promote innovative activity."

Market failure, as we address it, could be termed "technological or innovation market failure." Such market failure refers to the market—including both the R&D-investing producers of a technology and the users of the technology—underinvesting, from society's standpoint, in a particular technology. Such underinvestment occurs because conditions exist that prevent organizations from fully realizing or appropriating the benefits created by their investments.

The following explanation of market failure and the reasons for market failure follow closely from Arrow's seminal work, in which he identifies three sources of market failure related to knowledge-based innovative activity: "indivisibilities, inappropriability, and uncertainty" (Arrow 1962b, p. 609).[15]

Consider a marketable technology to be produced through an R&D process where conditions prevent the R&D-investing firm from fully appropriating the benefits from technological advancement. Other firms in the market or in related markets will realize some of the profits from the innovation, and consumers will typically place a higher value on a product than the price paid for it. The R&D-investing firm will then calculate, because of such conditions, that the marginal benefits it can receive from investment in such R&D will be less than could be earned in the absence of the conditions reducing the appropriated benefits of R&D below their potential—namely, the full social benefits. Thus, the R&D-investing firm might underinvest in R&D relative to what it would have chosen as its investment in the absence of the conditions. Stated alternatively, the R&D-investing firm might determine that its private rate of return is less than its private hurdle rate and therefore would not undertake socially valuable R&D.

This basic concept can be illustrated with Figure 1.1, which follows from Tassey (1997) and Jaffe (1998). The social rate of return is measured on the vertical axis, along with society's hurdle rate on investments in R&D. The private rate of return is measured on the horizontal axis, along with the private hurdle rate on R&D. A 45-degree line (dashed) is imposed on the figure under the assumption that the social rate of return from an R&D investment will at least equal the private rate of return from the same investment. Two separate R&D projects

Figure 1.1 Spillover Gap between Social and Private Rates of Return to R&D

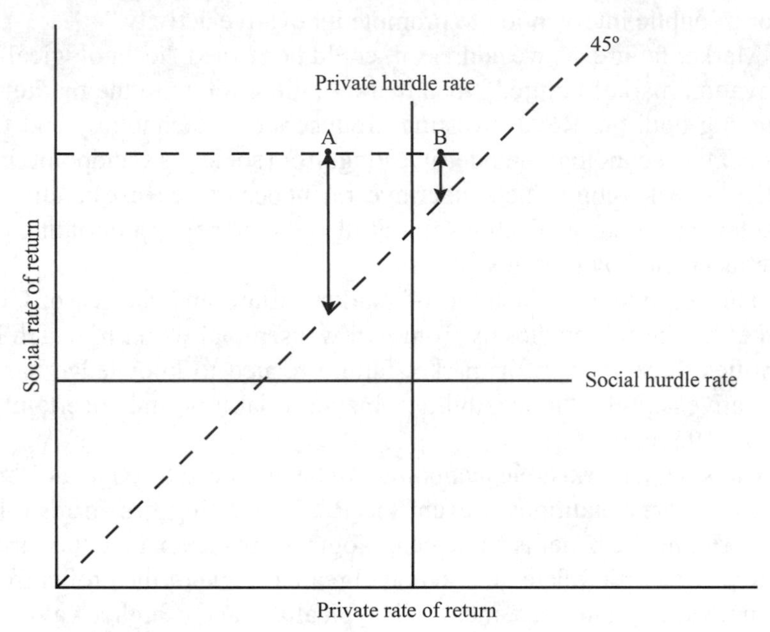

are labeled as Project A and Project B. Each is shown, for illustrative purposes only, with the same social rate of return.

For Project A, the private rate of return is less than the private hurdle rate because of barriers to innovation and technology (discussed below). As such, the private firm will not choose to invest in Project A, although the social benefits from undertaking Project A would be substantial (i.e., above the social hurdle rate).

The principle of market failure illustrated in the figure relates to appropriability of returns to investment. The vertical distance shown with the double arrow for Project A is called the spillover gap; it results from the additional value society would receive above what the private firm would receive if Project A were undertaken. What the firm would receive is less than its hurdle rate because the firm is unable to appropriate all of the returns that spill over to society. Project A is the type of project in which public resources should be invested to ensure that the project is undertaken.

In comparison, Project B yields the same social rate of return as Project A, but most of that return can be appropriated by the innovator, and the private rate of return is greater than the private hurdle rate. Hence, Project B is one for which the private sector has an incentive to invest on its own, even though the social rate of return is greater than the private rate of return. Or, alternatively stated, there is no economic justification for public resources being allocated to support Project B.

For projects of Type A, where significant spillovers occur, government's role has typically been to provide direct funding or technology infrastructure through public research institutions, which lowers the marginal cost of investment so that the marginal private rate of return exceeds the private hurdle rate.

Note that the private hurdle rate is greater than the social hurdle rate in the figure. This is primarily because of management's (and employees') risk aversion and issues related to the availability and cost of capital. These factors represent an additional source of market failure that is related to uncertainty. For example, because most private firms are risk-averse (i.e., the penalty from lower-than-expected returns is weighted more heavily than the benefits from greater-than-expected returns), they require a higher-hurdle rate of return compared to society as a whole, which is closer to being risk-neutral.[16]

To reduce market failures associated with inappropriability and uncertainty, government typically engages in activities to reduce technical and market risk (actual and perceived). There are several circumstances—termed barriers to technology and innovation—that cause market failure and an underinvestment in R&D.[17] Stated differently, there are a number of factors that explain why a firm will perceive that its expected private rate of return will fall below its hurdle rate, even when the social rate of return exceeds the social hurdle rate.[18] Individuals will differ about a listing of such factors, not only because they are not generally mutually exclusive, but also because of the relative importance of one factor compared to another in whichever taxonomy is chosen.

AN OVERVIEW OF THE BOOK

While the literature-based arguments supporting an innovation-based economic growth strategy are clear, many if not most of the empirical studies to date have narrowly focused on the aggregate relationship between R&D (or innovative activity) and productivity (or economic growth).[19] The more micro relationship between innovation and employment has largely been ignored by scholars to date, although in the current economic environment—that of a record federal deficit and unemployment only slightly below 10 percent—job growth is one issue that is raised in connection with nearly every public-sector expenditure.

In subsequent chapters we statistically examine employment growth associated with public support of R&D in small, entrepreneurial firms through the Small Business Innovation Research (SBIR) program. The SBIR program, which is discussed in detail in Chapter 2, is a public/private partnership that provides research grants to small firms to fund projects that are expected to result in commercialized innovations. Any government agency with an extramural research budget in excess of $100 million must set aside 2.5 percent of that budget to fund small-firm research that is aligned with the research mission of the agency. While the objective of the SBIR program is, generally, to stimulate innovation activity, it is reasonable to expect that employment growth at some level will follow from such public support of R&D. Our emphasis on employment growth is motivated not only by the academic literature but also by the current policy emphasis on job growth, especially as it relates to public support of innovation in small firms.

Our empirical analysis is based on information assembled by the National Research Council (NRC) of the National Academies in response to a congressional charge for it to make recommendations for changes in the SBIR program. As we discuss in Chapter 3, to meet this charge, the NRC assembled what is arguably the most complete database available for studying employment issues in innovative small firms; it is used for the empirical analyses presented in the following chapters.

As requested by Congress, the scope of the NRC database is limited to Phase II SBIR awards by the five largest agencies: the Department of Defense (DoD), the National Institutes of Health (NIH), the National

Aeronautics and Space Administration (NASA), the Department of Energy (DOE), and the National Science Foundation (NSF).[20] In Chapter 3, we give an overview of the history of the database, and we discuss the methodological approach used by the NRC to construct it. Firm-level and project-level variables in the data that are used in the statistical analysis in subsequent chapters are defined, and descriptive statistics on all variables are presented.

Chapter 4 is the first of three chapters specifically focused on employment impacts associated with SBIR funding. Its emphasis is on project-specific impacts, and the emphasis of Chapters 5 and 6 is on firm-wide impacts. That said, one might reasonably liken our empirical investigation into the innovation-employment relationship to a hunt for a Heffalump.[21] The reason for this is that previous research on the economic impact of R&D, entrepreneurship, and innovation on productivity and economic growth has been at an aggregate level. Whether an innovation-employment relationship will show up in the data at the project or even the firm level is an empirical issue, but logic suggests that the relationship might be there at least for some firms. A firm invests in R&D and, given time, that research results in a new technology.

The firm then commercializes the innovation. To meet the market demand for the innovation, assuming the firm plans to sell its innovation in the marketplace, the firm will likely hire new workers to produce it. However, if the firm innovates only with the intent of selling or licensing its innovation to others, job growth might not be present. Moreover, a lack of substantial job growth in the small firms conducting the R&D leading to innovation might even be expected, given that the theoretical foundation of public policy to stimulate innovation is based on market failure grounded in spillovers of value created by the innovating firms but appropriated by others. In any case, the innovation-employment relationship has yet to be studied systematically at any level of aggregation.

In Chapter 4, we identify at the SBIR project level variables that are correlated with the number of employees retained as a result of the technology developed during the SBIR-funded project. This descriptive analysis is narrow in its focus. It relates only to the direct, narrowly focused, and project-specific employment effects that occur from an SBIR project. Anticipating our findings, we know that SBIR-funded projects typically retain very few employees—on average, one to two employees—after the completion of the funded project to pursue the technology that was developed. In our findings, just over two-fifths of all completed projects retained no employees, and over one-third retained only one or two employees. Among many other findings, we see from the data that the number of retained employees is greater when the government actually uses the technology created with the SBIR award, thereby providing a clear commercial goal and ultimately creating a market for the commercialized result of the SBIR project.

Chapters 5 and 6, in contrast, focus on the broader, longer-term employment impact on the firm as a whole (as opposed to only the funded project) from SBIR funding. We argue that this broader focus is an appropriate next step for understanding more completely the employment impacts associated with these public investments in innovation. The analysis in Chapters 5 and 6 is based on an economic growth model suited to the data limitations in the NRC database, as discussed in Chapter 3. Our model, which is counterfactual in construction and implementation, allows us to make direct causal inferences about the impact of the SBIR-funded innovative activity on overall firm employment growth.

Specifically, in Chapter 5, we estimate the employment impact of SBIR funding on our sample of firms. We compare a firm's actual level of employment after receipt of an SBIR award and completion of the research project to the level of employment predicted by the firm's characteristics prior to receiving the award. In other words, our analysis focuses on a comparison of the firm's actual employment with the firm's predicted employment had the firm not received the SBIR funding for the sampled project. We find, on average among firms funded by the DoD, NIH, NASA, and DOE (but not by the NSF), that the difference between actual employment growth and predicted employment growth is positive. (The data in the NRC database relate to the year 2005.) For the DoD and NIH samples, the positive employment impact of a sampled award is statistically significant for several of the awards. However, the *average* difference between actual employment and predicted employment absent the award is not statistically different from zero for any of these agencies. We cannot conclude that, *on average*, SBIR awards generate a permanent long-run employment benefit to the firms funded by these agencies that are in our sample because, although there are average employment gains that appear quite substantial, the average gains are not statistically significant.

Despite the lack of statistical significance in the average employment gains for firms after receiving a sampled award, there are numerous individual cases of significant gains (statistically and economically) and large ranges in significance of the gains in the individual cases for all agencies. In Chapter 6, we build on those ranges in the results from Chapter 5 and posit an exploratory model to identify the variables related to SBIR-induced employment growth for the firm as a whole—that is, the variables related to the difference between actual employment for the firm and its predicted employment without the support of the SBIR award. We find, across agency projects, that there is considerable variation in the group of covariates associated with actual employment growth above predicted employment growth. The single covariate that is important, for SBIR-induced employment growth, across all agency samples is the presence of firm investments in intellectual property.

In Chapter 7, we explore the possibility of employment effects beyond the firm. Specifically, we examine descriptively the extent to which SBIR-funded projects result in commercial agreements with other U.S. firms or investors, comparing the extent of agreements made with foreign firms or investors. This inquiry builds on the empirically based discussion in Chapters 4 through 6 regarding such agreements. Our description shows that SBIR funds, and the associated technologies and employment growth, are benefiting not only U.S. firms and investors, but also foreign firms and investors through commercial agreements.

To our knowledge, our analysis is the first to use a counterfactual model to quantify the magnitude of the relationship between investments in innovation through public support of R&D and the resulting employment growth. As such, we view our empirical analysis of the employment growth in the small firms performing the R&D supported with SBIR funds as somewhat exploratory both in scope and in specification.

Our analyses in Chapters 5–7 represent an evaluation of one aspect of the SBIR program, namely employment growth, and such evaluations have attracted the administration's attention. Specifically, Peter Orszag, then the director of the Office of Management and Budget, wrote in an October 7, 2009, memorandum to the heads of executive departments and agencies on the subject of increased emphasis on program evaluation, "Rigorous . . . program evaluations can be a key resource in determining whether government programs are achieving their intended outcomes . . . and at the lowest cost. Evaluations . . . can help the Administration determine how to spend taxpayer dollars effectively and efficiently . . ." (Orszag 2009).[22]

Finally, in Chapter 8 we conclude with a brief summary of our findings, and we reiterate our view of their possible importance for future public policy. Our study of the NRC database reveals that the public's support of R&D in SBIR firms results at best in only modest employment gains within the SBIR firms themselves. However, our evidence supports the expectation that the accomplishments of these small firms will have employment effects that show up elsewhere in the U.S. economy and in foreign markets. Such employment effects beyond the SBIR firms themselves result from commercial agreements, such as the sale of technology rights to other firms, and through the spillovers of value from the R&D that the SBIR firms do not appropriate at all. As we have explained in this introductory chapter, such spillovers are at the heart of the market failure, which is the raison d'être for the public support of R&D in small, entrepreneurial firms.

Notes

1. The NBER dates the Great Depression as having lasted from August 1929 to March 1933, a period of 43 months. See http://www.nber.org/cycles/cyclesmain.html.
2. These unemployment statistics come from the Federal Reserve Bank of St. Louis. See http://research.stlouisfed.org/fred2/series/UNRATE.
3. The Council of Economic Advisers, as part of accountability and transparency provisions in ARRA, has released quarterly reports assessing the economic impact of ARRA.
4. The policy emphasis in ARRA on jobs, especially innovation-based jobs, can be traced back at least to the National Academy of Sciences report *Rising above the Gathering Storm: Energizing and Employing America for a Better Economic Future* (Committee on Prospering in the Global Economy 2007). This report was the foundation for the America COMPETES Act of 2007 (P.L. 110-69), which was reauthorized as the America COMPETES Reauthorization Act of 2010 in January 2011.
5. See Link (1987) and Link and Siegel (2003) for the literature related to culprits of the productivity slowdown.

6. R&E refers to research and experimentation activity. Experimentation is a more narrowly defined activity than development, according to National Science Foundation reporting definitions.
7. See Scott (2008) for a review of the literature about the history and effectiveness of the National Cooperative Research Act.
8. Hébert and Link (1988, 2009) provide an overview of the intellectual history of thought on who the entrepreneur is and what he or she does.
9. See, Hall, Mairesse, and Mohnen (2010) for a survey of this literature.
10. The Small Business Administration defines small firms as those having fewer than 500 employees. In the Acs and Audretsch (1990) study, innovations were defined by a contracted data-collecting firm as the number of new products reported in technology, engineering, and trade journals. Baldwin and Scott (1987) provide a review of the earlier literature about the relative innovativeness of small versus large firms.
11. The comparison does not necessarily mean that the smaller firms are more effective at producing innovations. Comparisons across firms of different sizes of their proportions of sales taken by R&D expenditures have been criticized because economies in large-scale use of R&D resources could imply that larger, less R&D-intensive firms are more effective at producing innovations than smaller, more R&D-intensive firms. Kohn and Scott (1982) establish the conditions for the possible relationships between the distributions of R&D across firm sizes and the distributions of their R&D output. Importantly, if R&D increased more than proportionately with firm size, then that would imply that the output of R&D activity also increased more than proportionately. However, the reverse relation does not hold. We do have evidence about the distribution of R&D output across firm sizes, however, and much of that supports the relative effectiveness of the smaller firms' R&D investments.
12. The ratio of R&D to firm sales was 15.8 percent for R&D firms with fewer than 25 employees and 3.1 percent for firms with more than 25,000 employees.
13. Were it not for Kahin and Hill (2010), we would not have known about this statement.
14. The conceptual importance of identifying market failure for policy is also emphasized, although without any operational guidance, in Office of Management and Budget (1996).
15. Although Arrow does not elaborate on indivisibilities and inappropriability in his paper, the concepts are well understood in the innovation literature. Recalling that Arrow (1962b, p. 609) defines innovation "as the production of knowledge," we know that the market does not price knowledge in discrete bundles, and thus because of such indivisibilities market prices may not send appropriate signals for economic units to make marginal decisions correctly.
16. There are two parts to the answer to the twin questions of how the social hurdle rate is determined and why it is represented as being less than the private hurdle rate. The first is grounded in the practice of evaluations, and the second is grounded in the theory of public policies to address market failure.

1) Regarding practice, the U.S. Office of Management and Budget (OMB) has mandated that a specified real rate of return be used as the rate for evaluation studies—that is, the rate to be considered the opportunity cost for the use of the public funds in the investment projects. The OMB (1992, p. 9) has stated that "constant-dollar benefit-cost analyses of proposed investments and regulations should report net present value and other outcomes determined using a real discount rate of 7 percent." That real rate of return (and the related nominal rates derived by accounting for expected inflation rates in various periods of analysis) has been far less than what the respondents in case studies have reported (Link and Scott 2011) as the private hurdle rate for comparable investment projects in industry during comparable time periods.

2) Regarding theory, when we evaluate public investment projects, we are invariably looking at cases where there has been some sort of market failure. To improve upon the market solution, the government has become involved (in a variety of ways, in practice) with an investment project. Just as market solutions for the prices of goods may not reflect the social costs for the goods (because of market failure stemming from market power, imperfect information, externalities, or public goods), the private hurdle rates that reflect market solutions for the price of funds—the opportunity cost of funds to the private firms—might not reflect the social cost of the funds. The government might decide that the appropriate social cost—the opportunity cost for the public funds to be invested—differs from the market solution. Typically, in practice, for the public R&D projects we have studied, the government evidently believes that it faces less risk than the private sector firms doing similar investments; hence it will believe a lower yield is satisfactory because the public is bearing less risk than the private sector firm going it alone with a similar investment. More generally, government must decide what the opportunity costs of its public funds will be in various uses, and in general that will not be the same as the market rate. However, all that having been said, clearly we know from Arrow's thinking about social choice that the government's decision about what the rate should be cannot possibly reflect the diversity of opinion in the private sector regarding the decision (Arrow 1963). Consequently, as a logical matter, one could not prove that the government's choice of the right hurdle rate is obviously correct, because diversity of opinion about the correct rate will not be reflected in the government's choice.

17. These factors are not independent of each other. They include high technical risk associated with the underlying R&D, high capital costs to perform the underlying R&D with high market risk, length of time to complete the R&D and commercialize the resulting technology, the fact that underlying R&D spills over to multiple markets and is not appropriable, that market success of the technology depends on technologies in different industries, that property rights cannot be assigned to the underlying R&D, that resulting technology must be compatible and interoperable with other technologies, and the high risk of opportunistic behavior when sharing information about the technology.
18. As Arrow (1962b) explains, investments in knowledge entail uncertainty of two types—technical and market. The technical and market results from technology

may be very poor, or perhaps considerably better than the expected outcome. Thus, a firm is justifiably concerned about the risk that its R&D investment will fail, technically or for any other reason. Or, if technically successful, the R&D investment output may not pass the market test for profitability. Furthermore, the firm's private expected return typically falls short of the expected social return, as previously discussed. This concept of downside risk is elaborated upon below and in Link and Scott (2001).

19. This global view is logical in the sense that aggregate economic growth is the overriding long-term objective of most public policies.
20. DOE, as opposed to the other abbreviations of departments here, has a capital "O" because of a departmental policy that capitalized it to distinguish the Department of Energy from the Department of Education.
21. A Heffalump is an imaginary elephant in the dreams of Piglet in the book *Winnie-the-Pooh*. The Heffalump gained academic notoriety through the writings of Kilby (1971), who wrote, "The search for the source of dynamic entrepreneurial performance has much in common with hunting the Heffalump . . . He has been hunted by many individuals using various ingenious trapping devices, but no one so far has succeeded in capturing him. All who claim to have caught sight of him report that he is enormous, but they disagree on his particularities . . . [But] the search goes on" (p. 1).
22. Hearings before the Senate Budget Committee were held on October 29, 2009, in support of Orszag's point of view. Also, two chapters about program evaluation have been included in the Fiscal Year 2011 and the Fiscal Year 2012 Budget of the U.S. Government; see http://whitehouse.gov/sites/default/files/omb/budget/fy2011/assets/spec.pdf and http://whitehouse.gov/sites/default/files/omb/budget/fy2012/assets/spec.pdf. Chapter 8 is on program evaluation and Chapter 9 is on benefit-cost analysis (OMB 2010).

2
Small Business Innovation Research Program

EVOLUTION OF A POLICY EMPHASIS ON INNOVATION IN SMALL FIRMS

As we mentioned in Chapter 1, productivity growth in the United States fell during the early 1970s and then again during the late 1970s and early 1980s, as it did in many industrialized nations. The evidence shows that total factor productivity growth during the late 1960s and early 1970s was less than one-half of that during previous decades.[1] While there have been many *ex post* explanations for the slowdown, such as the Organization of Petroleum Exporting Countries (OPEC) oil crisis in 1973 and industry's slow adjustment to it, there seems to have been a general agreement at that time and shortly thereafter that public policy aimed at stimulating innovation would be effective for stimulating economic growth. As a result, the Bayh-Dole Act was passed in 1980,[2] and the R&E tax credit was passed in 1981.[3]

In addition to the broad-based emphasis on innovation (i.e., the diffusion of patented technologies and the increase in R&D investments from these two legislative initiatives, respectively), a research report from the Massachusetts Institute of Technology's Neighborhood and Regional Change program was independently and coincidently published in 1979. Birch (1979, 1981) concludes in that report that three-fifths of the net new jobs between 1969 and 1976 were created by small firms with 20 or fewer employees. According to Birch (1979, p. 29), "On the average about 60 percent of all jobs in the U.S. are generated by firms with 20 or fewer employees, about 50 percent of all jobs are created by independent, small entrepreneurs. Large firms (those with over 500 employees) generate less than 15 percent of all net new jobs." Also, Birch (1979) reports that approximately 80 percent of net new jobs were created by firms with 100 or fewer employees.[4]

Birch's writings became the genesis for a new research field related to the economics of small businesses.[5]

Reflecting more broadly than on the productivity slowdown in the 1970s—a period of economic disequilibrium—Schultz (1980, p. 443) notes that "disequilibria are inevitable in [a] dynamic economy. These disequilibria cannot be eliminated by law, by public policy, and surely not by rhetoric. A modern dynamic economy would fall apart if not for the entrepreneurial actions of a wide array of human agents who reallocate their resources [to form new combinations] and thereby bring their part of the economy back into equilibrium."

It could be argued that this constant readjustment toward equilibrium by enterprises stimulates economic growth.[6]

One might think of the productivity recovery that began in the early 1980s in terms of entrepreneurial responses to disequilibria.[7] Or, using the terms of Audretsch and Thurik (2001, 2004), the recovery from the productivity slowdown reflects the end of the era of the "Managed Economy" (with predictable outputs coming from an established manufacturing sector) and the emergence of the "Entrepreneurial Economy."

In the Entrepreneurial Economy, characterized by the emergence of economic agents embodied with entrepreneurial capital, smaller firms have a greater ability to be innovative, or to adopt and adapt others' new technologies and ideas, and thus quickly and efficiently appropriate investments in new knowledge that are made externally.[8] Entrepreneurial capital engenders growth in new enterprises, and enterprise growth augments economic growth. Entrepreneurial capital also provides diversity among firms.[9] According to Audretsch and Thurik (2004, p. 144), with an emphasis on small firms within the Entrepreneurial Economy, "entrepreneurship has emerged [during the late 1970s] as the engine of economic and social development throughout the world." As a result, it is perhaps not surprising that policymakers toward the end of the 1970s embraced small firms as engines of future economic growth.

Carlsson (1992) offers two explanations for this shift in policy emphasis toward small, entrepreneurial firms. First, there had been a fundamental change in the world economy beginning in the mid-1970s.[10] Global competition was increasing, markets were becoming less fragmented, and the determinants of future economic growth were uncertain. Thus, entrepreneurial leadership adjusted to this disequilibrium. Second, and related, flexible automation was being introduced throughout the manufacturing sector, thus reducing economies of scale as a barrier for entry into many markets and thereby opening the door for smaller, entrepreneurial firms to enter and succeed.

It is not surprising, then, that the policy environment in the late 1970s and early 1980s was receptive to the establishment of the SBIR program.

THE CREATION OF THE SBIR PROGRAM

The program created as Small Business Innovation Research is a public/private partnership that provides grants to fund private-sector R&D projects. It aims to help fulfill the government's mission to enhance private-sector R&D and to commercialize the results of federal research.[11]

A prototype of the SBIR program began at the NSF in 1977 (Tibbetts 1999). At that time, the goal of the program was to encourage small businesses—increasingly recognized by the policy community as a source of innovation and employment in the U.S. economy—to participate in NSF-sponsored research, especially research with commercial potential. Because of the early success of the program at the NSF, Congress passed the Small Business Innovation Development Act of 1982 (P.L. 97-219; hereafter, the 1982 act).[12]

The 1982 act required all government departments and agencies with external research programs of greater than $100 million to establish their own SBIR program and to set aside funds equal to 0.20 percent of the external research budget.[13] In 1983, that set-aside totaled $45 million.

The 1982 act stated that the objectives of the program included the following four:

1) to stimulate technological innovation,
2) to use small business to meet federal research and development needs,
3) to foster and encourage participation by minority and disadvantaged persons in technological innovation, and
4) to increase private sector commercialization of innovations derived from federal research and development.

It is important to point out that employment growth was not a stated objective of the SBIR program in the 1982 act, or in subsequent reauthorizations.

As part of the 1982 act, SBIR program awards were structured and defined by three phases (Wessner 2004). Phase I awards were small, generally less than $50,000 for the six-month award period.[14] The purpose of Phase I awards is to assist businesses as they assess the feasibility of an idea's scientific and commercial potential in response to the funding agency's objectives.[15] Phase II awards were capped at $500,000; they generally last for two years. These awards are for the business to develop further its proposed research, ideally leading to a commercializable product, process, or service.[16] The Phase II awards of public funds for development are sometimes augmented by outside private funding (Wessner 2000). Further work on the projects launched through the SBIR program occurs in what is called Phase III, which does not involve SBIR funds.[17] At this stage, firms needing additional financing to ensure that the product, process, or service can move into the marketplace are expected to obtain it from sources other than the SBIR program.

As stated in the 1982 act, to be eligible for an SBIR award, the small business must have the following six characteristics:

1) It must be independently owned and operated.
2) It must not be the dominant firm in the field in which it is proposing to carry out SBIR projects.
3) It must be organized and operated for profit.
4) It must be the employer of no more than 500 employees, including employees of subsidiaries and affiliates.
5) It must be the primary source of employment for the project's principal investigator at the time of the award and during the period when the research is conducted.
6) It must be at least 51 percent owned by U.S. citizens or lawfully admitted permanent resident aliens.

In 1986, the 1982 act was extended through 1992 (P.L. 99-443). In 1992, the SBIR program was reauthorized again until 2000 through the Small Business Research and Development Enactment Act (P.L. 102-564). Under the 1982 act, the set-aside had increased to 1.25 percent;

the 1992 reauthorization raised that amount over time to 2.5 percent and reemphasized the commercialization intent of SBIR-funded technologies (see point 4 of the 1982 act, above).[18] The reauthorization also increased Phase I awards to $100,000 and Phase II awards to $750,000.[19] The 1992 reauthorization broadened objective 3, above, to focus also on women: "to provide for enhanced outreach efforts to increase the participation of . . . small businesses that are 51 percent owned and controlled by women."

The Small Business Reauthorization Act of 2000 (P.L. 106-554) extended the SBIR program until September 30, 2008. It retained the 2.5 percent set-aside and did not increase the amounts of Phase I and Phase II awards.[20]

Congress did not reauthorize the SBIR program by the legislated date of September 30, 2008; rather, Congress temporarily extended the program until March 20, 2009 (P.L. 110-235). The Senate version of the reauthorization bill (S. 3029) included among other things an increase in Phase I funding to $150,000 and an increase in Phase II funding to $1,000,000, with provisions for these funding guidelines to be exceeded by 50 percent. Also, the current 2.5 percent set-aside would increase to 3.5 percent at a rate of 0.1 percent per year over 10 years, except for the National Institutes of Health, which would stay at 2.5 percent. On March 19, 2009, the House and Senate reauthorized the SBIR program until July 31, 2009 (P.L. 110-10); it was again reauthorized through September 30, 2009, through a Senate continuing resolution (S. 1513). On September 23, 2009, a House bill (H.R. 3614) extended SBIR until October 31, 2009. A Senate bill (S. 1929) again extended the program until April 30, 2010; then another Senate bill (S. 3253) extended the program to July 31. Another House bill (H.R. 5849) extended the program to September 30, 2010; more recent bills (S. 3839 and H.R. 366) extended the program to January 31 and then May 31, 2011, respectively.[21] Still to be decided are whether the existing Phase I, Phase II, and Phase III processes should remain, whether the dollar amount of Phase I and Phase II awards should be changed, and whether venture capitalists should be involved in the SBIR process.[22] While this debate lingers in Congress, the Small Business Administration (SBA) on March 30, 2010, amended the SBIR policy directive to allow the threshold amount to increase to $150,000 for Phase I awards and to $1,000,000 for Phase II awards (as is proposed in the Senate version of the reauthorization bill). As of this writing, the program has been temporarily reauthorized until December 16, 2011 (H.R. 2112).

Eleven agencies currently participate in the SBIR program. Besides NASA, the NSF, the DoD, and the DOE, mentioned in Chapter 1, the other seven are the Environmental Protection Agency (EPA) and the departments of Agriculture (USDA), Commerce (DoC), Education (ED), Health and Human Services (HHS, particularly the NIH, which is also mentioned in Chapter 1), Transportation (DoT), and, most recently, Homeland Security (DHS). In 2005 (the year of the survey from which the data analyzed herein come, as discussed in Chapter 3), the DoD maintained the largest program, awarding about 51 percent of total dollars and funding about 57 percent of total awards in that year. Five agencies—the DoD, HHS, NASA, DOE, and NSF—account for nearly 97 percent of the program's expenditures, with the HHS (which includes the NIH) being the second most important, accounting for 30 percent of total dollars and 19 percent of awards in 2005 (Table 2.1). SBIR Phase II awards from these five agencies are studied in the subsequent chapters.

ECONOMIC ROLE OF THE SBIR PROGRAM

The general model of the spillover gap between social and private rates of return in Figure 1.1 in Chapter 1, which was introduced as a pedagogical device to support the argument for government's role in the innovation process, is expanded here in Figure 2.1 with specific reference to the SBIR program. For Project $A_{\text{without SBIR}}$ in Figure 2.1, the private rate of return is less than the private hurdle rate because of barriers to technology. As such, the private firm will not choose to invest in Project $A_{\text{without SBIR}}$, although the social benefits from undertaking the project would be great. The vertical distance measured by the distance from i_{social} to the 45-degree line at $i_{private}$ for Project $A_{\text{without SBIR}}$ is the spillover gap; it results from the additional value society would receive above what the private firm would receive if Project $A_{\text{without SBIR}}$ were undertaken. Project $A_{\text{without SBIR}}$ is precisely the type of project in which the public should invest, namely one in which the private sector would not invest because of market failure and one from which society would greatly benefit.

Table 2.1 SBIR Awards and Dollars, Fiscal Year 2005

Agency	Phase I awards	Phase I dollars	Phase II awards	Phase II dollars	Total awards	Total dollars
DoD	2,344	$213,482,152	998	$729,285,508	3,342	$942,767,660
HHS[a]	732	$149,584,038	369	$412,504,975	1,101	$562,089,013
DOE	259	$25,757,637	101	$77,852,565	360	$103,610,202
NASA	290	$20,183,648	139	$83,014,853	429	$103,198,501
NSF	152	$15,054,750	132	$64,101,179	284	$79,155,929
USDA	91	$7,195,211	40	$11,738,536	131	$18,933,747
DHS	62	$6,158,240	13	$10,241,202	75	$16,399,442
ED	22	$1,646,603	14	$6,749,980	36	$8,396,583
DoC	34	$2,373,433	19	$5,469,846	53	$7,843,279
EPA	38	$2,652,216	14	$3,540,251	52	$6,192,467
DoT	7	$679,154	3	$1,765,468	10	$2,444,622
Total	4,031	$444,767,082	1,842	$1,406,264,363	5,873	$1,851,031,445

[a] The National Institutes of Health (NIH) organization is under the Department of Health and Human Services (HHS).

SOURCE: Authors' tabulations using SBA data found at http://web.sba.gov/tech-net/public/dsp_search.cfm.

Figure 2.1 Spillover Gap between Social and Private Rates of Return to SBIR-Funded Research

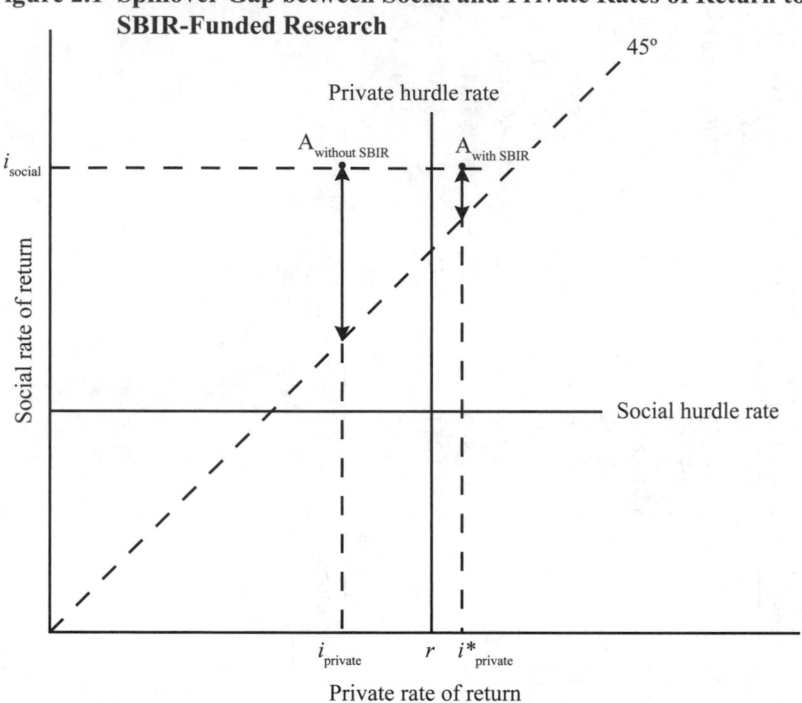

NOTE: *i** is the expected private rate of return with the SBIR support, while *i* is the expected private rate of return without the support.
SOURCE: Link and Scott (2010).

In Figure 2.2, we alternatively illustrate that reduction in risk in terms of a rightward shift in the distribution of the rate of return for the private firms.[23] The rightward shift of the distribution, and the concept of reducing the probability of returns to a likelihood lower than what would be acceptable to the private investors, applies equally well to the absolute level of net return (absolute return minus private investment) expected from the project. As shown in Figure 2.1, SBIR support increases the firm's expected private rate of return and thereby reduces the downside risk associated with undertaking R&D.[24] For each distribution—without SBIR funding (left distribution) and with SBIR funding (right distribution)—the expected rate of return is shown.[25] As drawn, with SBIR funding the expected private rate of return and the

Figure 2.2 Private Risk Reduction Resulting from SBIR Funding

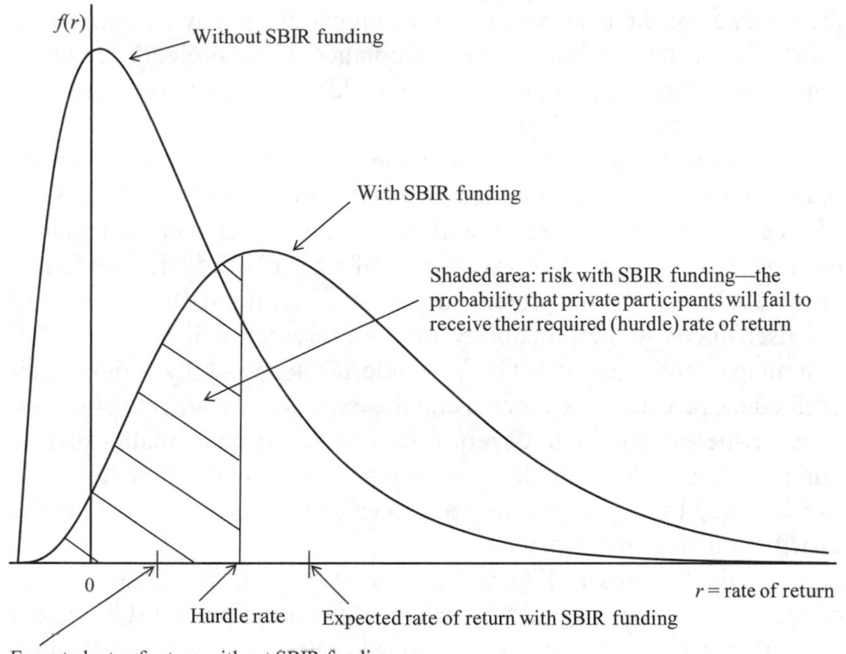

SOURCE: Link and Scott (2009).

variance in the private rate of return from the research project will increase. One can generalize that this will always be the case.[26]

Consider the left distribution—the distribution of the rate of return for the private firm without SBIR funding. As drawn, the private hurdle rate is to the right of the expected rate of return without SBIR funding, meaning that the private firm will not undertake this research because the firm will not receive its required rate of return. The risk of the project equals the area under this without-SBIR distribution that is to the left of the private hurdle rate. For those used to thinking of the variance of the distribution as the measure of risk, the downside risk—which is the probability of a rate of return less than the hurdle rate—might seem unusual. Variance measures the possibility that outcomes can differ from the expected outcome, while the downside risk measures the probability of an outcome departing to the downside of the hurdle rate. Note that the technical risk and the market risk for the project are reflected in the variance of the distribution: the technical goals may exceed or fall short of expectations, and market acceptance of the project's technical outcomes could do the same. The downside risk refers to the outcomes that fall short of the hurdle rate.

Now consider the distribution on the right in Figure 2.2—the distribution of the rate of return for the private firm *with* SBIR funding. With SBIR funding, the private firm will expect a return greater than its hurdle rate; thus the expected private rate of return with SBIR funding is drawn to the right of the private hurdle rate.[27] While SBIR funding will not itself increase the probability that the research will be successful, assuming hypothetically that it were undertaken absent SBIR funding, it will reduce private risk by increasing the expected private rate of return, because the expected rate of return will be based on a smaller private outlay.[28] Hence, SBIR funding leverages the private firm's investment, as illustrated by a greater expected return and a greater variance in the distribution, as explained above.

The shaded area in Figure 2.2 is what we call the downside risk of the project—that is, it is the probability that the project will yield a rate of return less than the private hurdle rate, even with SBIR funding. Hence, the amount of downside risk with SBIR funding is visually less than the downside risk associated with the research project in the without-SBIR funding case.

Although we will conclude that SBIR funding reduces risk, as defined operationally in terms of reducing the probability of a rate of return below the private hurdle rate, we emphasize that our analysis below is in no way wed to any particular measure of risk or any particular model of capital asset pricing with associated systematic and nonsystematic risk. Instead, our treatment encompasses any and all such models because the relevant risk, however it is perceived by private firms, is captured in the private hurdle rate, and the distributions of returns are otherwise represented by their expected values. In describing the effect of SBIR funding on the distribution of private rates of return, we are describing an underlying reality that would be reflected in the private hurdle rate—as determined by some model—and in the expected value of the returns.

EMPLOYMENT GROWTH AND THE SBIR PROGRAM

Although employment growth is not a stated objective of the SBIR program, it is reasonable to assume that future reauthorizations of the program will consider employment issues, explicitly or implicitly. We are of this opinion for three reasons, all of which were alluded to in Chapter 1.

First, the administration's view toward sustained economic growth is focused on job creation and employment growth, as reflected in the American Recovery and Reinvestment Act (ARRA) of 2009.

Second, while ARRA was legislated in the aftermath of what was the longest and most severe recession since the Great Depression, there are indicators that an emphasis on innovation-based employment growth will become an integral part, if not a centerpiece, of the implementation of the America COMPETES Reauthorization Act of 2010, signed into law in January 2011. Stine (2009, pp. 2–3) writes, "For the nation to maintain economic growth and a high standard of living, the United States must be competitive in a global economy. To be competitive, U.S. companies must engage in trade, retain market shares, and offer high quality products, processes, and services. Scientific and technological advances can further economic growth because they contribute to the creation of new goods, services, jobs, and capital . . ."

Stine (2009, p. 14) goes on to say that "science and engineering research is important to U.S. competitiveness because of its influence on U.S. economic growth . . . Additional federal funding of basic science and engineering research will make the nation more competitive by creating whole new industries and the related jobs, [as well as by] enhancing existing ones."

Continued support for science and engineering research is a major part of the 2010 budget (Sargent 2010), the 2011 budget, and likely of future budgets.

Third, the Office of Management and Budget is focusing on program evaluation as part of the administration's emphasis on accountability. Certainly, an economic evaluation of the SBIR program will be broader than an assessment of its enabling legislation's stated objectives, and it could focus on, among other things, job creation and employment growth as an economic benefit to be quantified.

Notes

1. Kendrick and Grossman (1980) report annual total factor productivity growth in the private business sector from 1948 to 1966 to have been 2.9 percent, but only 1.4 percent from 1966 to 1976. See Link (1987) and Link and Siegel (2003, 2007) for an overview of the productivity slowdown literature.
2. The University and Small Business Patent Procedures Act of 1980, also known as the Bayh-Dole Act, reformed federal patent policy by providing incentives for the diffusion of federally funded innovation results. Universities in particular were permitted to obtain titles to innovations developed with government funds. See Stevens (2004) for a historical account of the passage of the Bayh-Dole Act.
3. The Economic Recovery and Tax Act of 1981 (ERTA) included a tax credit for qualified R&E expenditures in excess of the average amount spent in previous years. See Atkinson (2007) and Tassey (2007) for an overview of the past effectiveness and current state of the R&E tax credit.
4. In subsequent research, Birch (1987) finds that small businesses with fewer than 20 employees accounted for 88 percent of all net new jobs over the 1981–1985 time period. Also, according to Birch (1981, p. 7), "Smaller businesses more than offset their higher failure rates with their capacity to start up and expand rapidly."
5. Some disagree with Birch's analyses and the findings of others who reached similar conclusions from analyses of firm and aggregated data. See, in particular, Davis, Haltiwanger, and Schuh (1996).
6. According to Klein (1979), the slowdown in productivity growth in the 1970s can be viewed in terms of a decline in businessmen's ability, or perhaps their desire, to deal with disequilibria.
7. Innovation and entrepreneurship are closely related concepts, as noted in Chapter 1. For Schumpeter (1934), the entrepreneur is the *persona causa* of economic development. The process of "creative destruction" is the essence of economic development and growth, writes Schumpeter (1950); it is a process defined by carrying out new combinations in production. An underlying hypothesis in the examination of employment growth in this book is that the Schumpeterian process of creative destruction is instigated in substantial part by small, entrepreneurial firms. The impact of the SBIR program on the process is explored.
8. Relatedly, the NSF and the SBA released the Gellman Report in 1976. It showed that small firms over the 1953–1973 period were more innovative per employee than were larger firms. This finding complements the more systematic research of Acs and Audretsch (1990) referenced in Chapter 1.
9. "It is the exchange of complementary knowledge across diverse firms and economic agents that yields an important return on new economic knowledge" (Thurik 2009, p. 10).
10. See Link and Tassey (1987) for documentation of this shift.
11. This section draws on Audretsch, Link, and Scott (2002); Link and Scott (2000); and Wessner (2000, 2004, 2008). For a taxonomy of public/private partnerships, see Link (1999, 2006).

12. The 1982 act amended the Small Business Act of 1953 (P.L. 85-536), which established the Small Business Administration.
13. SBIR is a set-aside program; it redirects existing R&D funds for competitive awards to small businesses rather than appropriating new monies for R&D. The 1982 act allowed for this percentage to increase over time.
14. The $50,000 amount and the $500,000 amount below are not stated in the 1982 act. They originally came from a policy directive to the Small Business Act of 1953, to which the 1982 act was an amendment. Both amounts are referenced explicitly in Senate Report 110-447 (U.S. Senate 2008).
15. "The objective of Phase I is to establish the technical merit and feasibility and potential for commercialization of the proposed R/R&D efforts and to determine the quality of performance of the small business awardee organization prior to providing further Federal support in Phase II." See http://grants.nih.gov/grants/funding/sbirsttr_programs.htm, p. 1.
16. "The objective of Phase II is to continue the R/R&D efforts initiated in Phase I. Funding is based on the results achieved in Phase I and the scientific and technical merit and commercial potential of the project proposed in Phase II." See http://grants.nih.gov/grants/funding/sbirsttr_programs.htm, p. 1.
17. "The objective of Phase III, where appropriate, is for the small business concern to pursue with non-SBIR funds the commercialization objectives resulting from the outcomes of the research or R&D funded in Phases I and II." See http://grants.nih.gov/grants/funding/SBIRContract/PHS2008-1.pdf, p. 1.
18. The percentage increased to 1.5 in 1993 and 1994, 2.0 in 1995, and 2.5 in 1997.
19. The reauthorization also stated that there should be an "adjustment of such amounts once every 5 years to reflect economic adjustments and economic considerations."
20. It is not uncommon for Phase II awards to exceed the $750,000 threshold.
21. On October 7, 2009, the House and Senate Armed Services Committees recommended that the DoD SBIR program be reauthorized for a year, until September 30, 2010. This recommendation became part of the DoD 2010 authorization.
22. See Schacht (2011) for a detailed overview of the reauthorization efforts.
23. This figure follows from Link and Scott (2001, 2010). Obviously, because we have shown a shift in a probability distribution, we have not thereby characterized a reduction in the most general notion of uncertainty following the distinction associated with Knight (1921). Alternatively, for those thinking of the more general notion of uncertainty in terms of subjective probability distributions for which different individuals might well have very different subjective prior beliefs, the figure could be said to illustrate a reduction in a more general notion of uncertainty than what is expressed by the frequency view of the distributions and risk.
24. We use a definition of risk that is focused on the operational concern with the downside outcomes for an investment. The shortfalls of the private expected outcomes from society's expected returns reflect appropriability problems. There are several related technological and market factors that will cause private firms to appropriate less return and to face greater risk than society faces. These factors underlie what Arrow (1962b) identifies as the nonexclusivity and public-good characteristics of investments in the creation of knowledge. The private firms' incomplete

appropriation of social returns in the context of technical and market risk can make risk in its operational sense unacceptably large for the private firm considering an investment. Operationally and with reference to Figures 2.1 and 2.2, Tassey (1992, 1997, 2003, 2005), for example, defines risk as the probability that a project's rate of return falls below a required, private rate of return or private hurdle rate (as opposed to simply deviating from an expected return). As illustrated in Link and Scott (2001), for many socially desirable investments, the private firm faces an unacceptably large probability of a rate of return that falls short of its private hurdle rate. Yet from society's perspective, the probability of a rate of return that is less than the social hurdle rate is sufficiently small that the project is still worthwhile.

25. Note that the expected rate of return does not necessarily correspond to the greatest frequency or probability density because the distribution of rates of return need not be symmetrical.
26. To capture the idea of limited liability for investors, we bound their return below by zero. Thus, the rate of return can be quite negative when the return falls below the amount invested, but because the return is bounded below at zero, the rate of return is bounded below by –100 percent. The expected private rate of return with SBIR support is: r = [return − (total project cost − SBIR funding)] / [total project cost − SBIR funding]. Let Z = (total project cost − SBIR funding). Then, r = (return − Z)/Z = [(return/Z) − 1]. The variance of r is: [$(1/Z)^2$ Var(return)], and it is a general proposition that as SBIR funding increases (and hence Z decreases) the variance in the private rate of return increases (since $(1/Z)$ gets larger). It is also a general proposition that the expected private rate of return, which equals E[(return/Z) − 1], must increase for the same reason. Furthermore, neither the expected social rate of return nor the variance in the social rate of return change at all. The social cost is the same, and the social return is the same.
27. SBIR funding need not affect the firm's private hurdle rate; that rate is set by corporate policy in most cases. Conceivably, because the operational measure of risk falls, the hurdle rate might fall as well in the presence of SBIR funding, and the stimulative effect of SBIR funding would hold *a fortiori*.
28. Also, in general it is possible that, apart from the funding itself, the support and guidance of the SBIR program will lower the probability of low returns.

3
National Research Council Database

BACKGROUND ON THE DATABASE

The Small Business Reauthorization Act of 2000 mandated that, among other things, the NRC conduct "an evaluation of the economic benefits achieved by the SBIR program" and make recommendations to Congress for improvements to the program. In its evaluation of the SBIR program, the NRC steering committee charged with the study took several approaches, including multiple surveys, interviews, and more than 100 case studies.[1] The results of the NRC's survey approach are described in this chapter.

The NRC conducted an extensive and balanced survey in 2005 based on a population of 11,214 projects completed from Phase II awards made between 1992 and 2001 by five agencies: 1) the DoD, 2) NIH (within the HHS), 3) NASA, 4) DOE, and 5) NSF. It was assumed as part of the NRC's sampling methodology that Phase II awards made in 2001 would be completed by 2005.

The five agencies surveyed accounted for nearly 97 percent of the program's expenditures in 2005, as was shown in Table 2.1 in Chapter 2. Table 3.1 shows the distribution of the population of 11,214 projects by funding agency, as well as the percentage of the 11,214 projects funded by each agency. The total number of projects surveyed from the 11,214 projects was 6,408.

The number and percentage of respondents from these 6,408 surveyed projects is shown in Table 3.2.[2] The total number of responding projects was 1,916, and the average response rate across all five agencies was 30 percent.

Also shown in Table 3.2 is the total number of projects in the final random sample of completed Phase II projects, by agency. For ease of reference, definitions of the variables from the survey and descriptive statistics are in the Glossary of Variables at the end of the book.[3]

Table 3.1 Population of SBIR Phase II Projects, 1992–2001

Agency	Completed Phase II projects	Percentage
DoD	5,650	50.38
NIH	2,497	22.27
NASA	1,488	13.27
DOE	808	7.21
NSF	771	6.88
All agencies	11,214	100.00

NOTE: "Percentage" column does not sum to 100.00 because of rounding.
SOURCE: Authors' calculations.

The NRC surveyed a number of nonrandomly selected projects because they were projects that had realized significant commercialization and the NRC wanted to be able to describe such interesting success stories. These nonrandomly selected projects are not considered herein. The data reduction process that was applied to each agency to arrive at the final random sample is in Table B.1 in the Technical Appendix, also found at the end of the book.

VARIABLES IN THE DATABASE

Three employment variables in the NRC database are the focal variables in the chapters that follow.[4] The variables, using 2005 data, include

1) number of employees,
2) number of employees hired as a result of the SBIR award, and
3) number of employees who, as a result of the award, were retained after the SBIR project was completed.[5]

Detailed information about types of employees is not available from the NRC survey, and this is not surprising. In small research firms, employees are technical workers, and they each wear many hats. As has been documented in previous case studies of Department of Defense SBIR projects (e.g., Wessner 2000), a given employee could be involved in research, marketing, and also management.

Table 3.2 Descriptive Statistics on the National Research Council Survey of Phase II Awards

Agency	Phase II sample size	Respondents	Response rate (%)	Random sample
DoD	3,055	920	30	891
NIH	1,678	496	30	495
NASA	779	181	23	177
DOE	439	157	36	154
NSF	457	162	35	161
All agencies	6,408	1,916	30	1,878

SOURCE: Authors' calculations.

The size of the SBIR awardee firms, as measured by the number of employees in the firm in the year of the survey, 2005, varies across firms and across agencies from a mean low of 41 employees among the 141 NSF firms in the random sample to a high of 63 employees among the 155 NASA firms. See the descriptive statistics in the Glossary of Variables. However, the range of firm sizes in 2005 is from 1 employee to 4,001.[6]

ISSUES OF SAMPLE SELECTION BIAS

The NRC database does not contain information on projects that were not funded by SBIR, and it does not contain information on projects for which a firm applied for SBIR support but was declined, or on comparable projects in firms that did not seek SBIR support. This lack of so-called matched pairs has been viewed by some (e.g., Wallsten 2000) as a long-standing shortcoming of SBIR data collected either by the NRC as part of its 2000 congressional mandate or at any other time, or by any of the agencies that participate in the SBIR program. However—and the NRC is aware of this—any effort to construct matched pairs would be flawed, and thus the NRC has not undertaken this task.

Firms that receive SBIR awards are small and unique in their research and organizational structure. Although Lerner (1999) has compared a large sample of SBIR awardees and matching firms, finding that the SBIR recipients have higher employment growth, Lerner and Kegler (2000, p. 321) explain that it is difficult with the matched pairs analysis "to disentangle whether the superior performance of the awardees is due to the selection of better firms or the positive impact of the awards."

Anticipating our analysis of possible employment growth in Chapter 5, our structural model provides a solution to the customary complaint about working with SBIR project data, namely the need for a control group. Our model provides a counterfactual control by its prediction of employment, based on preaward information only, to be compared to actual employment.

CROSS-AGENCY ANALYSES

In the chapters that follow, we examine, by agency, a number of employment issues using the NRC database. According to Wessner (2008, p. 109), "Comparisons between SBIR programs at different agencies . . . must be regarded with considerable caution . . . Widely differing agency missions have shaped the agency SBIR programs, focusing them on different objectives and on different mechanisms and approaches."

Throughout the remaining chapters, we compare our empirical findings across agencies, but we do not do so within the context of assessing or evaluating the relative performance of one agency's SBIR program against another's.

The primary goal of the DoD's SBIR program is the provision of technologies that are used as part of our nation's defense system, which is the mission of the DoD. In contrast, the NIH's primary mission for its SBIR program is the development of fundamental knowledge and its application for improving health. Pioneering the future of space exploration, scientific discovery, and aeronautical research is the mission of NASA and its SBIR program. The SBIR program within the DOE supports technical knowledge related to the agency's program areas. Finally, the National Science Foundation's SBIR program promotes science and the commercialization of science.[7] Table 3.3 provides a description of a randomly selected funded project for each agency.

We do, however, discuss cross-agency differences in the econometric results in the following chapters. By doing so, we are only emphasizing behavioral and descriptive differences in agency projects; we are certainly not advocating that our findings are a motivation for restructuring one agency's program based on results for another agency's program. Wessner's statement above about the caution needed when making comparisons, however, points out that *a priori* we do not expect to be able to make many generalizations—holding across all of the agencies' SBIR programs—about the factors determining the employment impact of SBIR awards.

CONDUCTING RESEARCH IN THE ABSENCE OF SBIR FUNDING

In Figure 2.2 in Chapter 2 we offered a theoretical rationale for the SBIR program. Namely, we concluded from that model of downside risk that in general the public sector should not fund a private-sector project with a private rate of return above the private hurdle rate even if the social rate of return is above the social hurdle rate.[8] The data in the NRC database allow us to provide some descriptive information in support of this argument.

As part of the NRC survey, each firm was asked, for each of its surveyed projects, "In the absence of this SBIR award, would your firm have undertaken this project?" Responses to this counterfactual question are summarized in Table 3.4, by agency. In every instance, over 50 percent of those surveyed responded "probably not" or "definitely not." And in every instance, less than 20 percent of those surveyed responded "probably yes" or "definitely yes." Also, such information is retrospective; it is not inconsistent with the policy prescriptions suggested by Figure 2.1 in Chapter 2, where the rates of return are the prospective expected rates of return. We view the responses in Table 3.4 as suggestive evidence in support of our behavioral model.

Table 3.3 Examples of SBIR Phase II Awards

Funding agency	Project description
DoD	In 2010, the DoD awarded $499,982 to SA Photonics, a California-based company, for a project titled "Analog to Information (A2I) Sensing for Software-Defined Receivers." Compressive sensing is based on the notion that the information content of a signal may be much less than its instantaneous bandwidth, and that this signal "sparseness" can be exploited directly during the sensing operation. When applied to the sampling of pulsed radar signals, compressive sensing can allow radar signals to be sampled below the Nyquist rate, allowing for a very small yet flexible software-defined Radar Warning Receiver (RWR). This compressive-based RWR can detect and parameterize unknown received radar signals over a frequency range from 0.1 to 20.0 GHz and can provide parameter estimation of the received power, carrier frequency, PRF, and pulse. With this information, available bandwidth can be determined. In this Phase II we will develop a prototype compressive-based RWR in order to validate its performance against simulated radar signals.
NIH	In 2009, the NIH awarded $518,708 to Edvotek Inc., a Maryland-based company, for a project titled "Science Education of Alcohol Metabolism." Human consumption of alcohol is ubiquitous. While it is reported that 90 percent of the population consumes various amounts of alcohol, only a small but sizable minority of the population abuses alcohol consumption. It is estimated that 10–20 percent of males and 3–10 percent of females develop persistent alcohol-related problems. Alcohol use by youth continues to be an important health focus for our nation. Alcohol misuse among adolescents is on the increase, and excessive drinking is associated with psychological, social, and physical harm to the individual, family, and society. During the Phase II award period, the company researched and tested new experiments and reagents in workshop and classroom settings prior to the marketing experiments for grades 7 to 12 on understanding how alcohol is metabolized.

Table 3.3 (continued)

Funding agency	Project description
NASA	In 2009, NASA awarded $599,996 in a Phase II project to Spectra Research Inc., an Ohio-based company, for a project titled "Dual Polarization Multi-Frequency Antenna Array." NASA employs various passive microwave and millimeter-wave instruments, such as spectral radiometers, for a wide variety of remote sensing applications, from measurements of the Earth's surface and atmosphere to cosmic background emissions. These instruments, such as the HIRAD (Hurricane Intensity Radiometer), SFMR (Stepped Frequency Microwave Radiometer), and LRR (Lightweight Rainfall Radiometer), provide unique data accumulation capabilities for observing sea-surface wind, temperature, and rainfall and significantly enhance the understanding and predictability of hurricane intensity. These microwave instruments require extremely efficient wideband or multiband antennas. For the Phase I SBIR program, Spectra Research Inc. teamed with scientists from the Georgia Tech Research Institute (GTRI) to apply new technological antenna advances and new antenna design tools toward solving the challenge of designing small, multifunction antennas that reduce the space, weight, and drag demand on the platform.
DOE	In 2003, the DOE awarded $749,845 to APS Technology Inc., a Connecticut-based company, for a project titled "Rotary Steerable Motor System for Deep Gas Drilling." This project will develop a new drilling tool that allows greater power to be delivered to the drill bit while allowing the drill bit to be continuously oriented in the desired direction, thereby eliminating "crooked" holes, which cause drilling, completion, and production problems.

(continued)

Table 3.3 (continued)

Funding agency	Project description
NSF	In 2010, the NSF awarded $471,495 to App2You Inc., a California-based company, for a project titled "Do-It-Yourself Database-Driven Web Applications from High-Level Specifications." This project will enable nonprogrammers to rapidly create and evolve fully custom-hosted, forms-driven workflow applications where users with different roles and rights interact. Such a platform will have a broad impact on organizations of all sizes by empowering nonprogrammer business-process owners to quickly and easily deploy applications that capture the business processes of their organizations. The platform has the maximum impact on enabling externally facing customer relationship management (CRM) for small and medium businesses (SMBs), which use the applications to facilitate and streamline interactions with customers and partners, achieve lower process-management and customer/partner servicing costs, increase customer/partner satisfaction, and grow revenues.

SOURCE: Web sites of the funding agencies.

Table 3.4 Responses to the Counterfactual Question, "In your opinion, in the absence of this SBIR award, would your firm have undertaken this project?"

	Percentage	Number
Answers from 593 DoD projects:		
Definitely yes	3.04	18
Probably yes	9.61	57
Uncertain	17.71	105
Probably not	32.72	194
Definitely not	36.93	219
Total	100.00	593
Answers from 338 NIH projects:		
Definitely yes	5.03	17
Probably yes	7.69	26
Uncertain	13.61	46
Probably not	27.81	94
Definitely not	45.86	155
Total	100.00	338
Answers from 111 NASA projects:		
Definitely yes	2.70	3
Probably yes	15.32	17
Uncertain	14.41	16
Probably not	35.14	39
Definitely not	32.43	36
Total	100.00	111
Answers from 114 DOE projects:		
Definitely yes	0.00	0
Probably yes	3.51	4
Uncertain	13.16	15
Probably not	44.74	51
Definitely not	38.60	44
Total	100.00	114
Answers from 121 NSF projects:		
Definitely yes	3.31	4
Probably yes	12.40	15
Uncertain	18.18	22
Probably not	42.15	51
Definitely not	23.97	29
Total	100.00	121

NOTE: Percentages may not sum to 100.00 because of rounding.
SOURCE: Authors' calculations.

PROJECT-SPECIFIC EMPLOYMENT GROWTH FROM SBIR AWARDS

As described above, the NRC database contains information on three employment variables. In Chapter 4, we focus narrowly on the third of these three variables, namely the number of employees in 2005 who, as a result of the award, were retained after the SBIR project was completed. We find that the project-specific employment effects are small. SBIR-funded projects in the NRC sample typically retain very few employees.[9] This result thus motivates subsequent chapters, wherein we investigate not project-specific effects but rather firm-wide effects and aggregate effects.

Notes

1. See Wessner (2008) for an overview of the findings from the case studies.
2. We thank Charles Wessner of the NRC for making these data available to us for this project.
3. As an aside, these descriptive statistics offer some interesting insights into the character of SBIR recipients and into the focus of their projects. While we do not discuss these points with reference to our analysis of employment growth, other than to hold them constant within our regression models, these points do provide some background about SBIR firms. First, more than half of all SBIR recipients are located in either the Northeast or the West, regardless of which was the funding agency. Second, when asked, "How did you (or do you expect to) commercialize your SBIR award?" about 40 percent of all firms that were funded by the DoD, NASA, DOE, or NSF responded "as hardware (final product, component, or intermediate hardware product)." Only firms funded by the NIH were more diversified: nearly 75 percent responded that their output was or is expected to be equally divided among hardware, software (i.e., for diagnostic purposes), or as a research tool. These details come from descriptive statistics that we developed and then presented in Table A.2.
4. The NRC did not systematically address employment issues as part of its mandated study. Rather, much of its focus was on case studies.
5. The exact wording of two relevant NRC survey questions is as follows: 1) "Number of current employees who were hired as a result of the technology developed during the Phase II project," and 2) "Number of current employees who were retained as a result of the technology developed during the Phase II project." As the mean values in Table A.2 suggest, the number of employees who were retained is sometimes greater than the number who were hired. As a result of the SBIR award

it could have been the case that x employees were hired but that y employees ($y > x$) were retained who would have been let go had it not been that the Phase II project was successful.

6. Obviously, a firm with 4,001 employees is not a small firm by definition from the SBIR guidelines—500 or fewer employees. However, this is the number of employees in this particular firm in 2005, which is obviously far greater than the number of employees in that firm in the year that it received its SBIR award. Hence, dramatic employment growth occurred in this particular case.
7. These missions and goals are elaborated on in Wessner (2008).
8. Of course, these rates of return are expected rates of return.
9. To belabor our elephantidae metaphor from Chapter 1, Heffalumps are rarely found on one's first hunting trip.

4
Project-Specific Employment Effects from SBIR Awards

DESCRIPTIVE STATISTICS ON RETAINED EMPLOYEES

We noted in the previous chapter that the NRC database contains information on three employment variables, by agency:

1) the 2005 number of employees in the firm, which we refer to in Chapters 5 and 6 as the actual number of employees in 2005;
2) the 2005 number of employees who were hired as a result of the technology developed during the funded Phase II SBIR project; and
3) the 2005 number of employees who were retained as a result of the technology developed during that same project.

Descriptive statistics on these variables, and all others, are in the Glossary of Variables.

Here we focus narrowly and exclusively on the third project-specific employment variable, employees retained by the firm as a result of the technology developed during the SBIR-funded project. In our opinion, focusing on retained employees is more relevant for a policy interpretation of our findings than focusing on hired employees, because retained employees reflect more permanent project-specific employment benefits from SBIR funding.[1]

As shown in Table 4.1, on average over two-fifths of all completed projects retained no employees after completion and over one-third retained only one or two employees. Thus, on average, the direct project-specific impact of SBIR-funded projects on employment is small, especially when compared to the mean number of employees in the firm in 2005, a number that ranged from 41 employees on average in NIH-funded firms to 63 employees on average in NASA-funded firms (see the descriptive statistics in the Glossary of Variables and Descriptive Statistics, Table A.1).[2]

Table 4.1 Percentage of Projects with Firms Retaining the Stated Numbers of Employees after Completion of Phase II Projects

Agency	*n*	0 employees	1–2 employees	3–4 employees	5–10 employees	> 10 employees
DoD	755	44.0	36.3	10.5	6.8	2.5
NIH	391	41.2	37.9	12.3	6.4	2.3
NASA	155	52.3	34.2	7.1	3.9	2.6
DOE	140	45.0	37.9	11.4	5.0	0.7
NSF	141	41.1	34.0	12.8	7.8	4.3

SOURCE: Authors' compilation from National Research Council database.

INTERPRETATION OF THE FINDINGS

Our interpretation of the patterns in Table 4.1 is, so to speak, the bottom line for the project-specific employment effects attributable to the SBIR award. SBIR-funded projects in this NRC sample typically retain very few employees—in absolute terms or as a percentage of the firm's total employment—to continue to pursue the technology developed in the funded project.

In the absence of a counterfactual model or a non-SBIR-funded sample of projects, we are unable to say from the data in Table 4.1 that SBIR-funded projects retain more employees than comparable non-SBIR-funded projects. However, when one takes a longer view and examines the employment effects on the firm as a whole (not just the employment effects on the funded project) within a counterfactual model, as we do for the next step, in Chapter 5, the empirical evidence still suggests that the employment impact of SBIR over time on the overall firm is, on average, small. Our model cannot reject the null hypothesis of no long-run employment effects on average across the randomly sampled SBIR projects. However, for four of the five agencies studied, the average effects on employment are positive and not insubstantial in their size. Also, there are a number of individual cases of statistically significant employment growth for firms winning SBIR awards as compared to the counterfactual prediction for employment in the absence of the award.

STATIC MODEL OF RETAINED EMPLOYEES

We explore here cross-project differences in the number of retained employees (*retainees*).[3] Because the number of retained employees associated with the project as of 2005 is a count variable, we estimate a negative binomial model with the expected incident rate, $rate_j$, for the number of retained employees in project j as being

$$(4.1) \quad rate_j = e^{\beta_0 + \beta_1 x_{1j} + \ldots + \beta_k x_{kj}}$$

for k explanatory variables. The number of years the project has existed—the project's age—is the exposure, E_j. Exposure is greater because with each year that passes since the Phase II project was funded there is the renewed chance the project will be seen as supporting a commercially viable expansion of the firm and additional permanent employees. Thus, the expected number of retained employees is

$$(4.2) \quad retainees_j = E_j e^{\beta_0 + \beta_1 x_{1j} + \ldots + \beta_k x_{kj}} = e^{\ln(E_j) + \beta_0 + \beta_1 x_{1j} + \ldots + \beta_k x_{kj}},$$

where *retainees* is the 2005 number of employees who were retained as a result of the technology developed during the Phase II project, and where exposure is included in the model as a natural logarithm with its coefficient constrained to be 1. The x's are firm and project characteristics related to the hypotheses below.

Our empirical analysis is descriptive; we are asking how the number of employees retained as a result of the technology developed during the Phase II project varies with the particular circumstances of the firm and the project.[4] Thus, our intent is to estimate cross-project differences in the number of employees who were retained as a result of Phase II–developed technology as a function of various explanatory variables. The explanatory variables considered in this econometric strategy are divided into eight groups, as categorized in the Glossary of Variables and Descriptive Statistics, Table A.1.[5]

Equation (4.2) was estimated as a negative binomial model, and the results are reported in Tables B.2 and B.3 in the Technical Appendix. Because the control for the commercial agreement variables substantially reduces the sample sizes, Table B.2 estimates the model for the larger samples available for each agency without those variables (hereafter referred to as the larger samples); Table B.3 then reestimates the model for the smaller samples for which the commercial agreement variables are available (hereafter referred to as the smaller samples). Standard errors are robust and are adjusted for clusters by firm because for some firms multiple Phase II SBIR projects are sampled.[6] The clustering allows for intragroup correlation in the errors for the multiple projects of a firm. Our interpretations of the regression results rely on calculated marginal effects.[7] Although exploratory, there are several interesting results in the estimated model.

To discuss the results, we begin with the SBIR research variables. Regarding whether a university was involved in the Phase II research (*univpartcptn*), the variable is a 0/1 qualitative variable, and the proportion of projects having university participation ranges widely over the samples, from a low of 0.26 for the DoD to a high of 0.53 for the NIH. Using the estimation of the larger samples, we see that university involvement has a significant effect on retained employees in only two cases. For the DoD, the significant effect is positive, while for the NIH, it is negative. Adding university participation to a DoD project is expected to increase the number of retained employees to 1.32 ($e^{0.275}$), or by 132 percent of the level in the absence of university involvement with the project. For the NIH, adding university participation decreases the expected number of retained employees to 0.646 ($e^{-0.437}$), or 64.6 percent of the level without university involvement. Perhaps this negative result for NIH projects with university participation reflects that those projects focus on contributions for which commercializable results can make use of the firms' existing production, marketing, and technology licensing staffs.[8] For the smaller samples in Table B.3, the same pattern exists. For the DoD sample the effect of university participation is positive and significant; for the NIH sample it is negative and significant.

The percentage of the firm's total R&D effort devoted to SBIR activities (*sbir-r&d-to-total-r&d*) has a significant effect on retained employees in three of the five agency samples. These effects are all positive, thus supporting the belief that commercialization potential—which requires a larger number of permanent employees—is greater for firms with more of their R&D devoted to SBIR projects. Increasing SBIR R&D support by a single percentage point is associated with an increase in the number of retained employees to 1.004 ($e^{0.00384}$), or by 100.4 percent of the previous level for the DoD, to 1.011, or 101.1 percent of the previous level for the NIH, and to 1.006, or 100.6 percent of the previous level for the NSF.

Typically, an SBIR award is about 40 percent of a firm's total R&D (see the descriptive statistics in the Glossary of Variables and Descriptive Statistics, Table A.1). Thus, the estimated effect is not insubstantial, because a single percentage-point increase from a base of about 40 percent is a small increase with a large effect. For example, a 10-percentage-point increase would be associated with an increase in retained employees to 1.039 ($e^{0.00384 \times 10}$), or 103.9 percent of the level without the increase. For the smaller samples, this pattern does not uniformly hold. In the DoD, NASA, DOE, and NSF samples the estimated impact is not significant. Only in the NIH sample is there a significant relationship, and that relationship is positive and of approximately the same magnitude as in the larger NIH sample.

Regarding the effect of the number of patents that resulted from the firm's SBIR-funded research (*sbirpatents*) on retained employees, the expected directional impact is ambiguous. Firms holding a large number of patents may already have personnel capable of developing such intellectual property, and thus may not have needed to retain additional employees to conduct the research on the current SBIR project. Following that argument, the estimated effect of patents on retained employees would be negative. For the larger samples, this finding is the case in the DoD, NIH, DOE, and NSF samples.

For the larger samples for the DoD, NIH, DOE, and NSF, the magnitude of the absolute value of the estimated coefficients on the number of SBIR patents is similar across agencies. While the estimated impact of this variable is statistically significant, it may not necessarily be economically significant. Across agencies, the mean value of the number of previous SBIR patents is 7.30. Thus, a one-unit increase represents on average a 14 percent increase in patents referenced to the mean. An increase of one SBIR patent reduces the expected number of retained employees to 99.4 percent ($e^{-0.0060}$) of the expected number of retained employees for the larger DoD sample. For the other three agencies, the percentages of expected number of retained employees are also close to 100 percent of what is expected without increasing the number of SBIR patents by one. After an increase of one patent, the percentage of retained employees expected, relative to the expected number without the change, is 98.5 percent for the NIH ($e^{-0.015}$), 98.0 percent for the DOE ($e^{-0.020}$), and 99.1 percent for the NSF ($e^{-0.0088}$). For the smaller samples, the estimated impacts are also typically statistically significant but again not very economically significant.

Contrary to the foregoing argument and evidence for a negative effect, the effect is positive in the NASA sample, and the effect is roughly the same absolute magnitude. Possibly, previous experience in patenting results from SBIR awards increases the marginal value of additional employees, resulting in the positive coefficient for that agency. Using the model estimated for the larger NASA sample, we calculate the marginal effect of an increase of one patent held by the firm from previous SBIR awards. Holding constant all the explanatory variables except for the number of previous SBIR patents, the expected number of retained employees given a one-unit increase in SBIR patents, relative to the expected number of retained employees without that change, is 1.00845 ($e^{0.00841}$)—an increase in expected retained employees of about 0.8 percent.

For the smaller samples, the same pattern is present, and the magnitude of the estimated coefficients is comparable. In any case, the effects here are not large.

The remaining SBIR research variable is the number of firm's previous Phase II awards that are related to the current Phase II project (*rldphII*). On the one hand, the need to retain additional employees from the current SBIR project might lessen as the number of complementary research projects the firm has undertaken grows greater, because the necessary employees might already be available within the firm. Thus, the estimated coefficient on this variable could be negative. On the other hand, through previously funded related research the firm might have built internal expertise in the form of human and technical capital that could, with a marginal employee, ensure successful commercialization from the current SBIR project. Thus, the estimated coefficient on this variable could be positive because the marginal value of an additional employee will be greater for the firm. In the larger samples, its estimated impact on retained employees is positive and significant in the DoD, DOE, and NSF samples, and negative and significant in the NIH sample. A one-unit change in the number of previous awards reduces the expected number of retained employees to 0.957, or 95.7 percent of what it would otherwise be for the NIH sample, and increases the expected number of retained employees to 1.17, or 117 percent of what it would otherwise be for the DoD sample, to 1.21, or 121 percent for the DOE sample, and to 1.32, or 132 percent for the NSF sample. At least based on the descriptive history of the employment experience for NSF awards, for example, it appears that a straightforward way to improve the employment trajectory for the NSF awards would be to focus them on firms with related Phase II experience. For the smaller samples, this same sign and significance pattern is present.

For the market variables, the result for the relationship between the presence of existing federal procurement (i.e., a federal system or acquisition program using the technology from the Phase II project so that *fedacq* equals 1.0—as is more often the case with DoD-funded projects than with those funded by other agencies) supports Nelson's observations. As Nelson (1982) observed in his case studies, such circumstances have historically been especially likely to foster commercializable results from privately performed, government-financed R&D.

Commercialization is especially likely because the government provides a market for commercialized results from the R&D, and also because the government has an active interest that can guide developments toward well-defined needs. Using the larger samples estimation, three of the five agencies show a significant effect for the existence of a federal acquisitions market, and all three are positive effects. The coefficients for the DoD, DOE, and NSF are 0.433, 0.594, and 0.837, respectively. For these agencies, when the federal government procurement is present, the numbers of retained employees are respectively 154 percent, 181 percent, and 231 percent of their levels without federal acquisitions programs using the Phase II technology. For the smaller samples, the same pattern exists, with the exception that the coefficient for NSF is nearly twice as large.

In the larger samples, none of the five agencies shows a significant effect for firms with a policy of a late evaluation of the commercial potential of the project (*lateeval*), although two of the effects are marginally significant and with different signs. On the one hand, a late evaluation of a project's commercial prospects allows better assessment of the commercial potential. On the other, the firm may be less likely to find such potential in the Phase II project's results. For the smaller samples, only in the NSF sample is the effect significant, and it is negative.

There are many significant differences in the number of retained employees for projects of different commercial types, and there are some significant differences across geographic regions. For example, in the larger NSF sample, projects in the Northeast are expected, other things being the same, to have 116 percent ($100 \times [e^{0.769} - 1]$) more retained employees than those in the West. The result in the smaller NSF sample is similar.

For both the larger and smaller samples, in every agency's sample at least one of the intellectual property variables—number of patents (*patents*), number of copyrights (*copyrights*), number of trademarks (*trademarks*), and number of publications (*publications*) from the technology developed from the Phase II project—is positive and significantly related to employee retention. Moreover, in only one case—the smaller NSF sample and copyrights—is there a significantly negative relationship between an intellectual property variable and retention. Patents, trademarks, and copyrights are private goods, and publications are a public good. More often than not, statistically significant effects on employees retained are observed when the project results in intellectual property, which with the exception of publications would be considered to be a private good rather than a public good. The intellectual property, when it has the attributes of a private good, is commercially valuable. The public good, publications, has a significant positive effect on retentions for just one of the five agencies' samples of projects. For NSF projects, at least, the presence of publications is a signal of commercially valuable intellectual property.

In the larger samples, the significant intellectual property effects are as follows. For the DoD, the mean number of patents is 0.91. Other things being the same, an increase of one patent is associated with an increase in expected retained employees to 1.03 ($e^{0.0307}$), or by 103 percent of the level otherwise. For the NSF, the effect is much larger; the mean number of patents is 1.04, and an additional patent increases the expected number of retained employees to 1.26 ($e^{0.230}$), or by 126 percent of the level without the extra patent. For the NIH, the mean number of trademarks is 0.40, ranging from 0 to 7 in the sample. These trademarks appear to have great commercial value; other things remaining constant, an additional trademark increases the expected number of retained employees to 1.46 ($e^{0.376}$), or by 146 percent of the level without the additional trademark. For NASA, the mean number of copyrights is 0.045, and the range in the sample is from 0 to 5. Copyrights are also associated with more retained employees. Other things being held constant, an additional copyright increases the expected number of employees to 1.51 ($e^{0.415}$), or by 151 percent of the level without the additional copyright. The mean number for the DOE is 0.143, and again the range in the sample is from 0 to 5. For the DOE sample of projects, an additional copyright increases the number of retained employees to 1.87 ($e^{0.626}$), or by 187 percent of the previous level. For the NSF, the mean number of publications is 1.70, and an additional publication increases the expected number of retained employees to 1.11 ($e^{0.106}$), or by 111 percent of the level without the additional publication. For the smaller samples, the sizes of these significant intellectual property effects are similar.

Turning to the funding variables, in the larger samples, non-SBIR federal funding is associated with a greater number of retained employees for all agencies; that positive effect is statistically significant for two and marginally so for one of the agencies. In the smaller NSF sample, the effect is negative, however, and significant. Both own-firm funding and personal funds also show positive effects on employee retention for all five agencies' projects in both the larger and smaller samples. For both of those funding variables in the larger samples, three of the five samples show effects that are statistically significant or at least marginally significant, and that is the case for two of the smaller samples. The only two significant positive effects for other domestic private-firm funding are present in both samples and are for the NIH and DOE.

In the larger samples, for U.S. private venture capital funding (*U.S.-private-venture-capital-to-total-investment*), foreign private funding (*foreign-private-investment-to-total-investment*), other private equity funding (*other-private-equity-investment-to-total-investment*), and state or local government funding (*state-or-local-government-funding-to-total-investment*), there are statistically significant effects with different signs in different agencies. College or university funding shows one large negative effect that is very marginally statistically significant. These cases where the statistically significant effects for funding variables have opposite signs reinforce the commonly held belief that each agency's SBIR program and projects have a distinct character. Hence, funding approaches that are associated with employee retention will differ across agencies. For the smaller samples, this same generalization applies.

Also, the results for the funding variables support another commonly held belief: that commercially successful projects are more likely when there is financial support from outside the SBIR program. Outside financial support signals that others who know about the commercial prospects are willing to invest. Also, the additional support is necessary for the Phase III success of a project. In both the larger and smaller samples, support for the importance of such outside finance is particularly strong in the DoD and NIH samples. For each of the five agencies, at least one of the variables for outside finance had a statistically significant positive association with retained employees.

To provide perspective on variables with the differing signs for significant effects and also to illustrate the magnitude of the effects, consider the ratio of additional development funding during the Phase II project from foreign private investment to total investment in the Phase II project (*foreign-private-investment-to-total-investment*). On the one hand, foreign investors could eventually retain any developed technology and commercialize and/or use it overseas. In that case, it could be that the greater the extent of foreign funding, the less likely the awarded firm would incur, or would be allowed by its investors to incur, the cost of hiring and retaining additional employees. Thus, the estimated coefficient on this variable could be negative. On the other hand, if the intent of foreign investors is to transfer overseas any technology developed from the current SBIR project, then they might hire and retain employees to ensure that the research project is in fact successful and to codify the basis for the developed technology. Moreover, they may well foresee a U.S. production presence. Thus, the estimated coefficient could be positive.

In the larger samples, the estimated coefficients on foreign investments are positive for the DoD, NIH, and DOE projects and negative for the NASA and NSF projects. The marginal impact of an increase of 10 percent of total funding (say, from 0.05 to 0.15) in foreign investments on the expected number of retained employees as a proportion of the expected number of employees retained without the increase in foreign private investment is the base to the natural logarithms raised to the power of one-tenth of the estimated foreign investment coefficient. Thus, for the DoD, NIH, and DOE, the effect for expected retained employees from increasing the percentage of the foreign private funding by 10 percent is respectively 1.314 (a 31.4 percent increase), 1.342 (a 34.2 percent increase), and 1.384 (a 38.4 percent increase) of what the expectation would be without the change in foreign private investment. For NASA and the NSF, the effect is respectively 0.636 (a 36.4 percent decrease) and 0.797 (a 20.3 percent decrease) of what would occur without the change in funding. The effects for the smaller samples are broadly similar.

For another example, consider two internal funding variables, the ratio of additional development funding during the Phase II project from the SBIR firm's own investments (*own-firm-funding-to-total-investment*) and the ratio of additional funding from the principal's own investments (*personal-funds-to-total-investment*). We expect asymmetry of information about the potential success of a Phase II research project. When own-firm or personal funds are invested, reflecting the "inside" information of the firm and its principal, the probability may be especially high for success of the research project and thus the hiring and retention of additional employees. Therefore, the estimated coefficients on own-firm and personal funding should be positive.

For the larger samples, in all of the agency analyses, the estimated coefficients on own-firm and personal funding are positive, and most are also significant. Whenever the estimated coefficient on either of these variables is significant, the magnitude of the estimated coefficient is greater than the estimated coefficient on any of the other funding variables. The significant effects are as follows. For an increase in own-firm funding by 10 percent of the total funding, for the DoD, NIH, DOE, and NSF, respectively, the effect for expected retained employees is 1.123 (a 12.3 percent increase), 1.383 (a 38.3 percent increase), 1.168 (a 16.8 percent increase), and 1.156 (a 15.6 percent increase) of what the expectation would be without the change in own-firm funding. For an increase in personal funds by 10 percent of the total funding, for the DoD, NIH, NASA, and NSF, respectively, the effect for expected retained employees is 1.166 (a 16.6 percent increase), 1.192 (a 19.2 percent increase), 2.80 (a 180 percent increase), and 1.376 (a 37.6 percent increase) of what the expectation would be without the change in personal funding. Again, for the smaller samples, the effects are broadly similar.

Information was available on various commercial agreements only for the smaller samples, as discussed above. The results in Table B.3 show that commercial agreements are often associated with statistically significant differences for retained employees. Commercial agreements have both positive and negative effects on the retention of project-specific employees. One effect that is especially pronounced comes from the sale of technology rights. In all agencies except for the DoD, the relationship between the sale of technology rights and retained employees is negative, and it is significant in the NIH, NASA, and NSF samples. Simply, when technology rights are sold, the firm no longer retains employees for the project to which the rights belong. But, as we discuss in Chapter 7, the sale of technology rights is not that prevalent, and it is almost nonexistent to foreign firms or investors. All of the statistically significant findings for the small samples are summarized in Table B.4 in the Technical Appendix.

CONCLUSIONS

In summary, the descriptive statistics in Table 4.1 show that the impact of SBIR funding on employment that is specifically associated with the SBIR project is typically small. Although there are a few cases where a large number of employees were retained because of their SBIR Phase II awards, on average the numbers of retained project employees have been small. What impact there is varies systematically across agencies based on distinct project and firm characteristics.

While there are cross-agency sample differences in variables that are related to the level of project employees retained as a result of the technology developed during Phase II, several general patterns are clear. (These patterns are summarized in Table B.4 in the Technical Appendix.) For example, the following patterns hold true:

- Federal acquisition is positively associated with the retention level, other things being the same.
- Projects producing educational materials have fewer retained employees.

- Intellectual property is consistently positively associated with retained employees for the projects of all five agencies.
- Funding above and beyond the funding provided by the SBIR program is consistently positively associated with retained employees for projects awarded by the DoD, NIH, and DOE; it is consistently negatively associated with retained employees for the NSF's projects.
- Many types of commercial agreements are associated with fewer retained employees for at least some of the agencies' SBIR projects, supporting the hypothesis that the agreements allow the SBIR firm to transfer the production—and hence the employment supporting that production—to other firms, using the technology developed in the Phase II project. An exception is marketing/distribution agreements, which are associated with greater retention, other things being the same.

Notes

1. Unfortunately, the NRC data are not sufficient to track employees hired into a funded project and then retained after the project is completed.
2. Importantly, the employment impact is also small when compared to the employment in the firm when it applied for the Phase II project award. On average, employees in the firm at that time ranged from 21 employees for the NIH sample to 47 employees for the NASA sample.
3. Throughout the book we make reference to the mnemonic variable name in parentheses to guide the reader through the statistical tables in the Technical Appendix.
4. We are looking at the sample of projects for descriptive patterns in the data for the respondents and do not control for selection in the estimation of Equation (4.2). The number of employees retained is essentially a commercialization decision, and elsewhere (Link and Scott 2009, 2010) we have shown that the commercialization decision is uncorrelated with selection for all but one of the five agencies' samples of projects. For that one agency's sample of projects, the error in the model of response is correlated with the error in equation for the commercialization model. The effect of selection in the error of the commercialization equation is then of course important for prediction of commercialization conditional on response or in the population for that one agency. However, even in that one case, with or without control for selection, the estimates of the coefficients on the explanatory variables are essentially the same. It is those coefficients that are important here; therefore, we use the simpler specification rather than the more complex simultaneous es-

timation of the negative binomial model for *retainees* and the probit model for selection into the sample by response.

5. There are eight groups of variables in these tables; *retainees* is the only firm characteristic variable used in the analysis in this chapter.
6. We depart in Tables B.2 and B.3 from the traditional notation for significance at the 0.01, 0.05, and 0.10 levels. Our prior is that the profession is poorly oriented toward conveying important *p*-value information, especially in exploratory empirical work.
7. To describe marginal effects we use the incident rate ratio because it places the effect in the perspective of the outcome without the effect. From Equation (4.2), holding constant all variables except x_i, given a change in x_i, the expected number of retained employees relative to the expected number without that change is the base for the natural logarithms raised to the power of the product of the coefficient for x_i and the change in x_i. For a unit change in x_i, the expression is called the incident rate ratio and is simply the base for the natural logarithms raised to the power equal to the coefficient for variable x_i.
8. The probability of commercialization among the selected projects is actually somewhat higher for the NIH projects than for the DoD projects. See Link and Scott (2010, Table 11, p. 600).

5
Quantifying Long-Run Employment Growth from SBIR Funding

We described in Chapter 4 the project-specific employment effects attributable to SBIR funding in terms of the number of employees retained as a result of technology developed during the funded Phase II project. The mean number of retained employees ranges from one to two across funding agencies; however, on average, over two-fifths of all completed projects retain no employees and over one-third retain only one or two employees. To our knowledge, these measures are the first to document the SBIR project-specific employment effects. In fact, to our knowledge, these findings are the first to quantify the project-specific employment effects from any public R&D source. Thus, there are no other studies of the employment impact of public support of innovation in small firms to which to compare these project-level findings. Future research will certainly reinvestigate this topic at the project level.

In this chapter we investigate descriptively and econometrically the broader, longer-run impacts of SBIR funding on the *overall* employment growth of SBIR grant-recipient firms (as opposed to the employment growth of the recipient firm's funded research project). In our view, this broader measure is the more important indicator through which to glean insight about the employment impact of public support of innovation.

One methodology to evaluate the employment effects attributable to SBIR is to assign the awards randomly among the applicants submitting acceptable proposals. Another approach could be to find firms that were equally worthy of the awards, except that because of limited resources funds were only available for a subset, and the agency randomly decided which firms would receive them. Alternatively, firms could be ranked according to their qualifications for the award and compared to those below the cutoff point. These possibilities are moot because the NCR database only contains firms that received awards.

We compare the firm's actual level of employment after receipt of an SBIR award and completion of the research project to the level of employment predicted by the firm's characteristics prior to the award.

Specifically, our analysis focuses on a comparison of the firm's actual 2005 employment—after SBIR funding for the sampled Phase II project—with the firm's predicted 2005 employment had the firm not received the SBIR funding for the sampled project. We refer to this as a measure of SBIR-induced employment growth.

Our analysis in this chapter is dynamic in the sense that it takes a broad and general view of the impact of an SBIR award on a firm's employment trajectory over time, as opposed to the narrow look taken in Chapter 4 at project-specific employment gains associated with SBIR funding. The funding for the sampled project might, or might not, have enabled the firm to meet challenges and thrive at a critical juncture in its history; without the award those challenges might not have been met successfully. Thus, the firm having received the award might have been able then to retain employees to work not only on the funded project but also on other research activities unrelated to the project.

Moreover, our analysis is innovative in that it follows from a formal counterfactual model; this model is estimated taking into account response bias and thus addresses criticism about the lack of the comparative potential of the NRC database as discussed in Chapter 3.

Econometrically, we use available information about firms prior to their SBIR award to predict what their employment performance would have been in 2005 (the year of the NRC survey and thus the last year of available data) absent the award. We then compare actual 2005 employment after the award was received to that prediction to see if the award significantly changed employment growth or performance in the firms.[1]

COUNTERFACTUAL MODEL OF EMPLOYMENT GROWTH

Consider a general continuous-time growth model; apart from random error, the firm grows continuously.[2] The number of the firm's employees, y, at time t is

$$(5.1)\quad y(t) = ae^{gt}e^{\varepsilon},$$

where $y(t)$ is the number of employees t years after the firm was founded; a, an estimated parameter of the model, is the start-up number of employees; g is the annual rate of growth of employees; and ε is a random error term. The growth rate of employment is a function of various explanatory variables, x_1 to x_k, as

$$(5.2)\quad (\partial y(t)/\partial t)\,/\,y(t) = g = b_0 + b_1x_1 + \ldots + b_kx_k\,.$$

We refer to Equation (5.2) as our growth model. Below we hypothesize particular explanatory variables to include in the model.

From Equation (5.1), it follows that

$$(5.3)\quad \ln y(t) = \ln a + gt + \varepsilon\,.$$

Substituting Equation (5.2) for g in Equation (5.3) yields

$$(5.4)\quad \ln y(t) = \ln a + b_0t + b_1x_1t + \ldots + b_kx_kt + \varepsilon\,.$$

Equation (5.4), which we refer to as our employment model, is estimated using preaward firm information, and then $\ln y(t)$ is predicted for 2005 (with preaward firm characteristics and allowing the firm to grow from its founding until 2005), and that prediction is compared to the survey data on actual postaward employees in 2005.[3] Stated differently, the prediction of firm employment in 2005 is a prediction that is based on information about the past before the Phase II project began, and thus it is a prediction not using any information about the funded project.

We ask if having the Phase II award (i.e., receiving SBIR research funding) changed the trajectory for the firm that is predicted by the preaward history using only firm information that does not reflect any knowledge about the Phase II project. Thus, the numerical difference between the firm's actual employment in 2005 with the SBIR award and the firm's predicted level of employment based on preaward information used in the estimation of Equation (5.4) is the employment impact of SBIR funding on the firm.

Given the data limitations of the NRC database as discussed in Chapter 3, and especially having data on firm employment at only two points in time—1) when the Phase II award was made t years after the firm was founded and 2) in 2005—the credibility of our approach of course depends on the accuracy of the prediction from Equation

(5.4), and the accuracy of the prediction from Equation (5.4) of course depends on the accuracy of the specification of the growth model, *g*, in Equation (5.2).

To increase the robustness of our model and thus to provide a degree of confidence about the general predictive power of Equation (5.4), we allow for predictions that not only vary across firms but also vary through time by age for each firm.

In the following section we offer two hypotheses regarding variables to include in Equation (5.2), and we tie each variable to basic economic theory or to the extant academic literature.[4]

TESTABLE HYPOTHESES

Hypothesis 1

The growth rate of employment, *g*, will eventually decrease with firm age, all else remaining constant.

This hypothesis follows from the general concepts of diminishing returns and learning by doing (Arrow 1962a), and from the empirical research of Evans (1987). Following Arrow (1962a), the economic returns the firm can earn over time from its resources will eventually decline. The firm, or more specifically the firm's entrepreneur, will learn over time to optimize, but unless the physical capital or the entrepreneurial capital of the firm changes, the impact of this learning by doing will diminish. Recall that the average age of a funded firm, across agencies, is between 9 and 13 years, and it is likely that in some of the earlier years the firm's revenues were $0 (see the descriptive statistics for the variables, Table A.2 in the appendix).

Hypothesis 1 could also be viewed as an extension of the empirical work of Evans (1987), who demonstrated that among larger and older manufacturing firms the growth rate of employment decreased with firm age.

Hypothesis 2

As a complement, and perhaps as an alternative to Hypothesis 1 relating growth rate of employment to the age of the firm, research-oriented firms especially may benefit from prior market and research experience and avoid any downturn in the growth rate of employment.

The more experienced the firm with research—such as, for example, any SBIR research prior to the SBIR award being studied for the firm in our particular sample—the greater the likelihood that the firm will avoid a decline in its employment growth. Of course, research experience can be controlled, and, given that control, the nonlinear effect of the firm's age might still be expected. In that sense, this second hypothesis is a complement to the first. But it can also be considered as a potential alternative hypothesis here, because age itself will measure experience for the firm, which may allow it to avoid any decline in the growth rate of its employment.

Hypothesis 2 follows from the argument that with prior experience, and with the learning by doing that has occurred, the entrepreneur has an increased base of tacit and codified knowledge to draw upon to meet the inherent vagaries of research and the inevitable pressures of competition, all else remaining constant.

This follows from Jovanovic (1982, p. 649): "Firms learn about their efficiency as they operate in the industry. The efficient grow and survive; the inefficient decline and fail." Thus, each firm in an industry has a level of efficiency, and that efficiency, assumed here to be correlated with experience, helps to ensure that it survives and grows. (For the estimation of Equation [5.4], which follows from Equation [5.2], we are observing firms that have survived for t years.)

ECONOMETRIC RESULTS

In the growth model in Equation (5.2), we quantify the x's with the following variables.[5] Firm age (*firmage*), its square, and its cube are included in the growth model to allow for an unconstrained estimation of the age effect on the growth rate of employment.[6] Therefore, Equation (5.4) contains the variable t, along with the firm age variable multiplied by t, firm age squared multiplied by t, and firm age cubed multiplied by t. In effect, Equation (5.4) has as regressors t, t^2, t^3, and t^4.

From Hypothesis 1, the growth rate of employment is, eventually, a decreasing function of the age of the firm. Importantly, when we use our estimated employment growth model in Equation (5.4) to project the firm's level of employment in 2005 in the counterfactual case that the firm had not received the SBIR award, the predicted annual rate of employment growth for the firm not only differs across firms, but also it changes for each firm as the firm ages.

Regarding the firm's previous market and research experience, the following variables were considered: a qualitative variable denoting whether the firm was founded at least in part as a result of the SBIR program (*sbirfnd*), the number of other firms started by the firm's founder(s) (*otherstarts*), and the number of previous Phase II awards (*prevphII*).[7] From Hypothesis 2, such experience is positively related to the growth rate in employment.

Also included in Equation (5.2) are controls for the firm's innate innovative capabilities. Specifically, we control for the business (*busfndrs*) and academic (*acadfndrs*) background of the firm's founder(s), and for how well the firm has evaluated the commercial potential of its SBIR projects during or after the completion of the Phase II research (*lateeval*).

Our argument about the importance of the origins of the firm is based on several strands of literature. Sociologists contend that individuals imprint their beliefs and values on situations in which they have control.[8] The founding history or background of the firm's founders establishes blueprints or organizational forms that direct, to varying degrees over time, the observed behavior of the organization that they influence.[9] From an economic perspective, if the firm's capabilities are influenced, if not determined, by the founder's background, then that background should influence how the firm achieves its comparative advantage.[10] And, from a management perspective, the resource-based view of the firm posits that firms are bundles of heterogeneous resources and capabilities, and that those resources and capabilities influence the firm's competitive strategy. To the extent, then, that these bundles of heterogeneous resources and capabilities have been influenced by the entrepreneur's blueprint for the firm and attendant skills within the firm, aspects of the firm's or founder's background are related to the firm's performance.[11] Related to the firm's evaluation of the commercial potential of its SBIR projects, comparative advantage and chosen competitive strategy characterize the firm's "dynamic capabilities," in the phrase of Augier and Teece (2007). We conjecture that employment growth is a manifestation of the firm's dynamic capabilities.[12]

Finally, variables are included for the geographic region in which the firm operates (*northeast*, *south*, and *midwest*, with the effect for *west* left in the intercept) to control for regional differences in labor markets, in competitive environment, and in the availability of venture capital and other sources of external financing to assist the firm's commercialization efforts.[13]

Equation (5.4) is estimated with control for response to the project survey; that is, it is estimated as a maximum likelihood model with selection. Thus, Equation (5.4) is estimated simultaneously with a model of the probability of response to the project survey. The response model ideally captures the underlying, ultimate explanations for nonresponse by firms for which awards were sampled. Those ultimate explanations include not only behavioral stories about decisions to respond or not, but also include structural factors that are correlated with a lack of response to the NRC survey because a firm no longer has valid postal or e-mail addresses.

One project-specific control included in the probability of response model is the amount of the Phase II award (*awardamt*). *A priori*, we hypothesize that the amount of the Phase II award will have a positive effect on the probability of response because firms receiving larger awards might be more inclined to respond as a quid pro quo for the greater SBIR support. Moreover, the larger awards may be associated with firms that have stable information about how to contact them.

The firm's number of Phase II awards that were among the population of 1992–2001 projects for sampling in the NRC survey (*phIItsvy*) is also held constant in the response model.[14] The variable is one measure of firm size; *a priori*, firms that are larger in this dimension are hypothesized to be more likely to respond to the survey because they have the resources available to do so and, perhaps as well, as a quid pro quo for greater SBIR support. Again, the variable may also be associated with stable contact information increasing the likelihood of successfully getting a response for sampled projects.

Also included as a regressor is the number of a firm's Phase II projects that were surveyed by the NRC (*numsvyd*). On the one hand, this variable might have a negative effect on the probability of response if a larger reporting burden lowers the probability of response. On the other hand, firms with larger numbers of surveyed projects are those that have received more SBIR projects, so they might be inclined to respond because of that fact. Furthermore, such firms might have the resources at hand to respond more readily to the survey than the firms with fewer awards, and, again, the variable is likely to be correlated with the type of institutional stability that avoids a failure of the survey to find its recipient.

Finally, we hypothesize that firms would be less likely to respond to a specific project survey as the project grew older, because institutional memory about the project may have faded with time or the older set of facts simply may have been more difficult, and hence costly, to assemble and report. Thus, we expect that the project's age (*prjage*) will be negatively related to the probability of response. Also, other things being the same, such projects could be correlated with a loss of contact information needed to successfully reach the recipient of the survey.

See Table B.5 in the Technical Appendix for descriptive statistics on the variables in the response model, for each agency.

Equation (5.4) is estimated with robust standard errors to allow for heteroskedasticity in the errors in the equation. Also, the standard errors are clustered by firm to control for correlation of the errors in the equation for the observations of those firms with more than one Phase II project in the sample.[15] The estimation uses probability weights, also called sampling weights. The sampling weights for the reporting categories for each agency are defined and provided in Table B.1 in the Technical Appendix. The estimated results from Equation (5.4), with control for response to the survey, are in Table B.6 in the Technical Appendix. They show cross-agency differences in the coefficients' sizes, signs, and levels of significance for the variables in our employment growth model, Equation (5.4), as discussed below.[16]

INTERPRETATION OF THE FINDINGS

Regarding Hypothesis 1, for all five agencies the empirical results support the eventual decline of employment growth with the age of the firm. To illustrate graphically the nonlinear relationship between employment and age, we calculated g using only pre–Phase II award information.[17] Our calculations of g, by agency, are illustrated in Figure 5.1.

We interpret our econometric results and our illustrative calculations of g as support for Hypothesis 1. The DoD example is the most striking, with an increase in the growth rate of employment to approximately $t = 20$, followed by the hypothesized eventual decline to approximately $t = 65$ (see the note to Figure 5.1). The decline in the growth rate of employment begins immediately for the other four agencies. (See Table B.6 in the Technical Appendix for details about which variables contribute to the reduction in growth rate as t increases.) We interpret this graphical support for Hypothesis 1 as empirical evidence of the credibility of our growth model in Equation (5.2) and thus of our employment model in Equation (5.4).

Empirical support for Hypothesis 2 is mixed across agencies. Among DoD, NIH, DOE, and NSF firms, at least one measure of previous market and research experience is positive and significant (*sbirfnd*, *otherstarts*, *prevphII*). But one of the three experience variables has a negative and significant estimated coefficient among NIH, NASA, and NSF firms, contrary to our hypothesis. Only among NSF firms is there empirical evidence that the number of other firms started by the firm's founder(s) is positively related to employment growth.

Finally, considering the remaining control variables, across all agencies a dimension of the founders' backgrounds—the imprint of their beliefs and values—is positively related to employment growth (*busfndrs*, *acadfndrs*).

The survey information about actual firm employment at the time of the sampled award and actual firm employment in 2005 can be used to calculate the actual annual growth rate of employees after the firm receives its Phase II award, as follows:[18]

$$(5.5) \quad g_{post} = [\ln(empt^{A}_{05}) - \ln(empt)] \,/\, tsa\ ,$$

where g_{post} is the postaward growth rate, $empt^{A}_{05}$ is the actual level of employment in 2005, *empt* is the number of employees at the time of the award (previously defined), and *tsa* is the amount of time since the Phase II award (*tsa* = 2005 − year of the Phase II award; it is therefore equivalent to *prjage*).

Table 5.1 shows the mean growth rate of firms, by agency. Across all agencies, the mean annual growth rate of postaward employees was over 8 percent, and by agency the range of mean growth rates was

Figure 5.1 Calculation of *g* from Equation (5.2), by Agency

Panel A

A company's estimated annual growth rate, *g*, as a function of its age, *t*, at the time of the DoD Phase II award

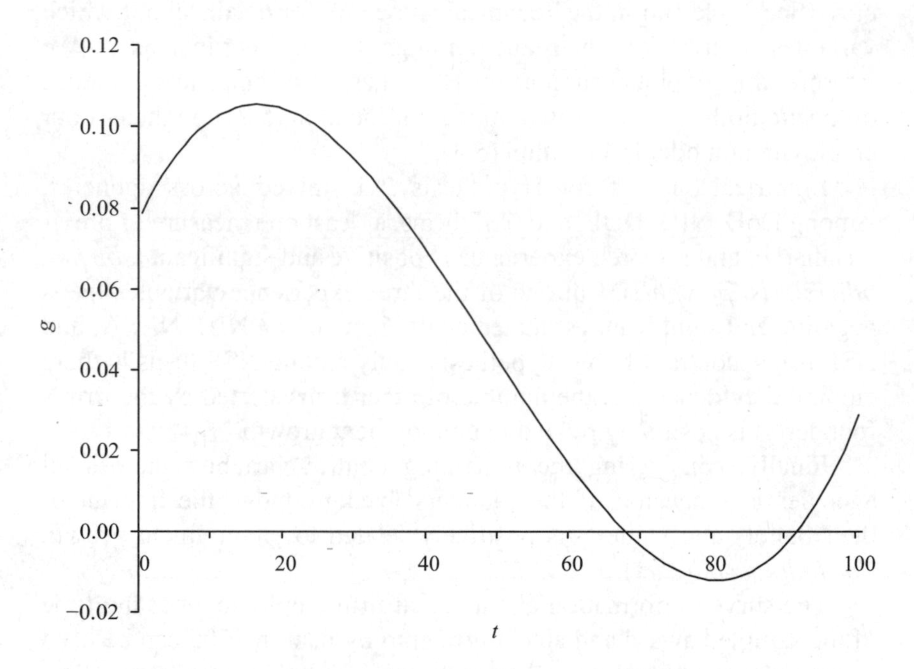

NOTE: Only preaward information is used to estimate the model. Graph charts the estimated annual growth rate *g* as a function of the company's age, *t*, as of the time of the Phase II award for a company with *sbirfnd* = 0, *otherstarts* = 0, *busfnders* = 1, *acadfnders* = 1, *prevphII* = 8.55 (the mean for *prevphII* in the 755-observation sample), *lateeval* = 1, and *south* = 1. Note that in the DoD 755-observation sample, the age of the companies at the time of the Phase II award ranged from 0 to 98 years. However, there were only two companies with age greater than 65 years; one was 93 years old and the other was 98 years old, so the negative portion of the graph of the predicted growth rates does not correspond to any actual companies in the data. Rather it is the portion of the curve needed to fit the data showing growth rates for the two especially old companies.

Figure 5.1 (continued)

Panel B

A company's estimated annual growth rate, *g*, as a function of its age, *t*, at the time of the NIH Phase II award, for companies 0 to 37 years old

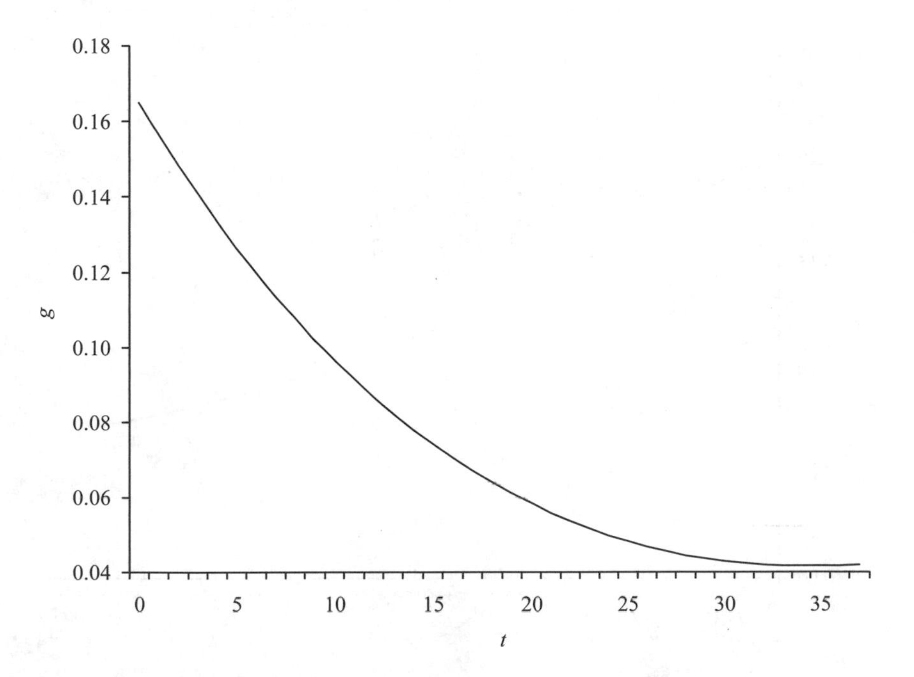

NOTE: Only preaward information is used to estimate the model. Graph charts the estimated annual growth rate *g* as a function of the company's age, *t*, as of the time of the Phase II award for a company with *sbirfnd* = 0, *otherstarts* = 0, *busfnders* = 1, *acadfnders* = 1, *prevphII* = 2.12 (the mean for *prevphII* in the 391-observation sample), *lateeval* = 0, and *south* = 1. Note that in the NIH 391-observation sample, the age of the companies at the time of the Phase II award ranged from 0 to 100. However, there were only two companies among the 391 that were older than 37 years old.

Figure 5.1 (continued)

Panel C

A company's estimated annual growth rate, *g*, as a function of its age, *t*, at the time of the NASA Phase II award

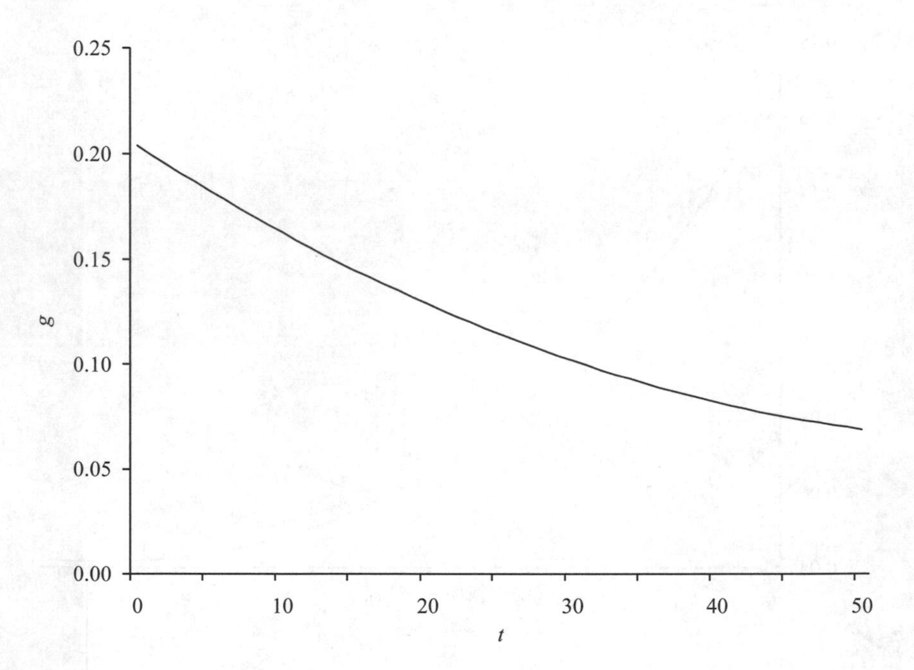

NOTE: Only preaward information is used to estimate the model. Graph charts the estimated annual growth rate g as a function of the company's age, *t*, as of the time of the Phase II award for a company with *sbirfnd* = 0, *otherstarts* = 0, *busfnders* = 1, *acadfnders* = 1, *prevphII* = 13.75 (the mean for *prevphII* in the 155-observation sample), *lateeval* = 1, and *south* = 1. Note that in the NASA 155-observation sample, the age of the companies at the time of the Phase II award ranged from 0 to 50 years.

Figure 5.1 (continued)

Panel D

A company's estimated annual growth rate, *g*, as a
function of its age, *t*, at the time of the DOE Phase II award

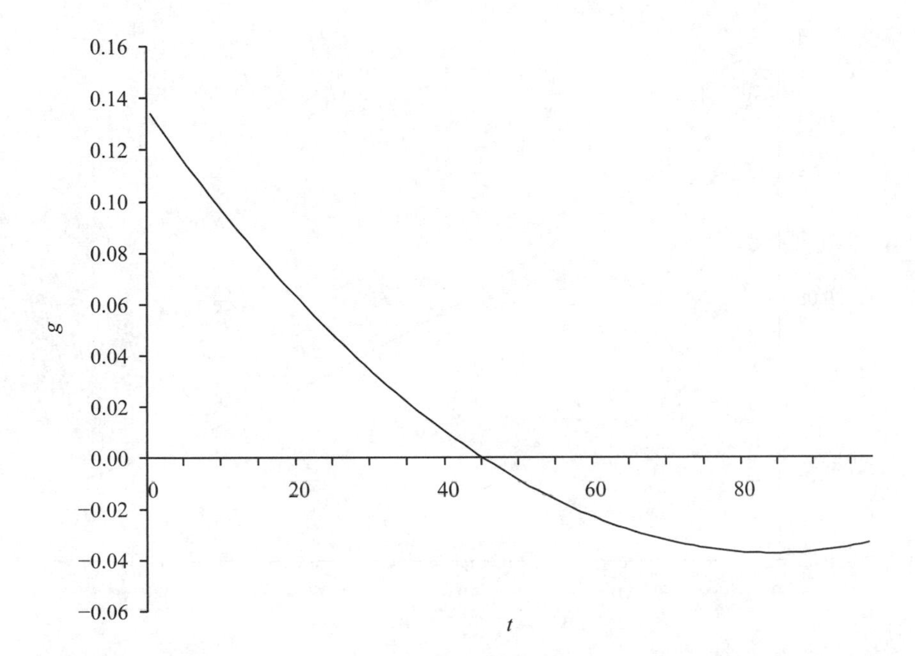

NOTE: Only preaward information is used to estimate the model. Graph charts the estimated annual growth rate *g* as a function of the company's age, *t*, as of the time of the Phase II award for a company with *sbirfnd* = 0, *otherstarts* = 0, *busfnders* = 1, *acadfnders* = 1, *prevphII* = 7.11 (the mean for *prevphII* in the 140-observation sample), *lateeval* = 1, and *south* = 1. Note that in the DOE 140-observation sample, the age of the companies at the time of the Phase II award ranged from 0 to 50 years for all but one of the companies. The one company with age greater than 50 years was 97 years old.

Figure 5.1 (continued)

Panel E

A company's estimated annual growth rate, g, as a function of its age, t, at the time of the NSF Phase II award

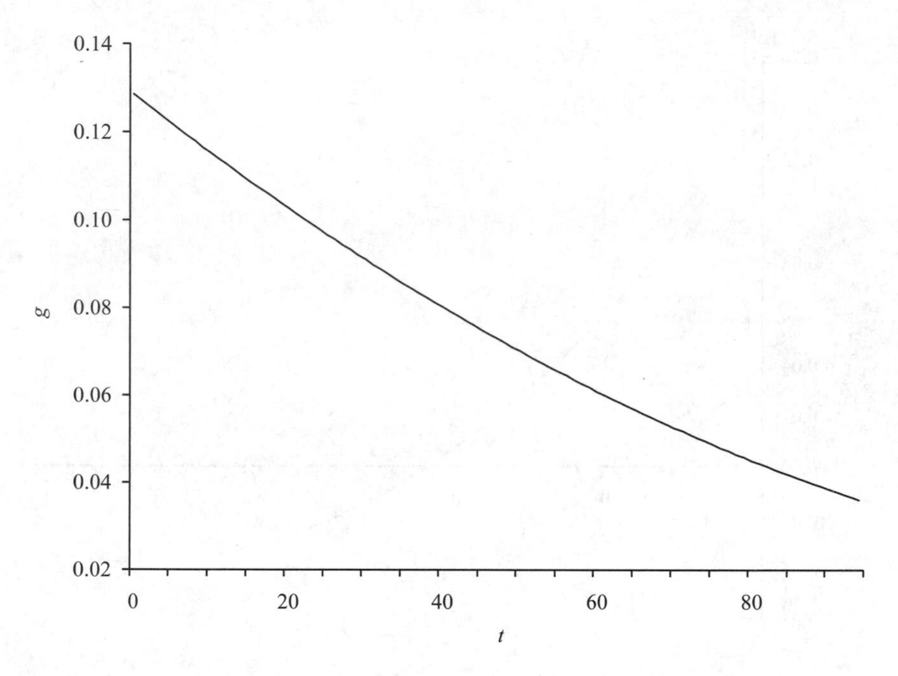

NOTE: Only preaward information is used to estimate the model. Graph charts the estimated annual growth rate g as a function of the company's age, t, as of the time of the Phase II award for a company with *sbirfnd* = 0, *otherstarts* = 0, *busfnders* = 1, *acadfnders* = 1, *prevphII* = 4.35 (the mean for *prevphII* in the 141-observation sample), *lateeval* = 1, and *south* = 1. Note that in the NSF 141-observation sample, the age of the companies at the time of the Phase II award ranged from 0 to 50 years for all but two of the companies. One company with age greater than 50 years was 92, and the other was 94 years old.

SOURCE: Authors' calculations from Equation (5.2).

from 5.9 to 9.2 percent. Over 75 percent of the firms had a positive growth rate, by agency, with nearly 40 percent of the firms enjoying growth greater than the agency mean. For about 10 percent of the firms, employment growth after receipt of the SBIR award was exceptional (growth rates greater than one standard deviation above the mean), and those firms averaged nearly 40 percent growth in employment (with a range from 31 percent for DOE firms to nearly 45 percent for NASA firms).

Table 5.2 shows the mean level of actual (i.e., reported on the survey) employment in 2005 and the mean level of firm employment predicted (when *t* is redefined so that the firm has existed right up to 2005) from the estimated results in Table B.6 in the Technical Appendix, by agency. Here, "predicted" refers to the predictions of employment conditional on the response to the survey, because, of course, we have the actual employment outcomes only for those firms that responded to the survey and completed the questionnaires.[19]

The information in Table 5.2 suggests that among firms in four agencies—DoD, NIH, NASA, and DOE—there was, on average, positive employment growth that is attributable to the SBIR award. Although SBIR-induced employment growth is on average substantial, none of these average gains is statistically significant. As shown in Table 5.3, the difference between the natural logarithms of actual (*A*) employment in 2005 and predicted (*P*) employment, or

$$diff = [\ln(empt^{A}_{05}) - \ln(empt^{P}_{05})] ,$$

is, at the level of the individual projects' effects on their sponsoring firms' employment, on the whole not statistically different from 0 in any of these four agencies. For the NSF, where the average induced growth was negative, none of the individual firm cases for overall firm employment growth induced by the SBIR award is significantly different from zero. However, there are many individual cases for the DoD (10 cases) and NIH (16 cases) samples for which the positive employment growth attributable to the SBIR award is statistically significant at the 5 percent level for a two-tailed test. Moreover, there is a wide range of calculated outcomes from negative to positive, and we explore the determinants of the differences in firm employment growth attributable to the SBIR awards in the following chapter.

Table 5.1 Actual Growth Rate of Employees after the Firm Received a Phase II Award

Agency	Number of firms	Mean growth rate (%)	Number, percentage (in parens), and mean growth rate of firms with a nonnegative growth rate	Number, percentage (in parens), and mean growth rate of firms with a growth rate greater than the mean growth rate	Number, percentage (in parens), and mean growth rate of firms with a growth rate greater than 1 standard deviation above the mean growth rate	Number, percentage (in parens), and mean growth rate of firms with a growth rate greater than 2 standard deviations above the mean growth rate
DoD	755	9.2	629 (83)	294 (39)	84 (11)	35 (4.6)
			12.6	22.4	40.1	53.8
NIH	391	8.2	323 (83)	164 (42)	39 (10)	11 (2.8)
			11.7	19.5	36.5	57.0
NASA	155	7.3	123 (79)	59 (38)	12 (8)	5 (3.0)
			11.2	20.1	44.8	64.8
DOE	140	5.9	106 (76)	60 (43)	15 (11)	7 (5.0)
			9.7	15.1	30.6	39.4
NSF	141	6.7	110 (78)	59 (42)	14 (10)	5 (4.0)
			10.7	17.8	37.8	55.9

SOURCE: Authors' calculations.

Table 5.2 Mean Actual ($empt^{A}_{05}$) and Predicted ($empt^{P}_{05}$) Employment in 2005 for Phase II SBIR Award Recipients, by Agency

Agency	*n*	Actual	Predicted
DoD	755	59.93	31.38
NIH	391	60.41	19.10
NASA	155	62.54	48.71
DOE	140	55.71	34.11
NSF	141	40.99	79.16
N	1,582		

SOURCE: Authors' calculations.

Table 5.3 Ratio of the Logarithmic Metric, *diff*, for SBIR-Induced Employment Growth to the Standard Error of the Forecast

Range for ratio	DoD	NIH	NASA	DOE	NSF
≥ 2.0	10	16	1	1	0
≥ 1.5 and < 2.0	25	12	0	3	0
≥ 1.0 and < 1.5	42	29	5	11	0
≥ 0.0 and < 1.0	349	139	79	46	61
≥ −1.0 and < 0.0	272	153	61	58	64
≥ −1.5 and < −1.0	39	28	8	18	10
≥ −2.0 and < −1.5	17	9	0	1	6
< −2.0	1	5	1	2	0
N	755	391	155	140	141

NOTE: Alternatively, using the usual z-value of 2 to leave a probability of approximately 0.025 in each tail of the distribution, calculating the 95 percent confidence interval around each observation's predicted employment as the antilogs of the predicted ln(employment in 2005) plus or minus the antilogs of two standard errors of the forecast, we would of course have the number of cases for which the actual 2005 employment fell outside the confidence interval, as shown in the first and last rows of the table here, combined: 11 cases for DoD, 21 cases for NIH, 2 cases for NASA, 3 cases for DOE, and 0 cases for NSF. Stated differently, although there are in absolute number several cases for DoD and NIH for which actual employment is significantly greater than the employment predicted by the counterfactual model, on the whole, the performance of the companies after winning the sampled Phase II award is not typically significantly different from what would have been predicted without the award. There is, nonetheless, quite a range of differences between actual and predicted employment, and we shall explore the determinants of those differences. Furthermore, as seen in Tables 5.2 and 5.4, the average difference between actual and predicted employment is positive for DoD, NIH, NASA, and DOE, although those average differences are not statistically significant.

SOURCE: Authors' calculations.

Although on the whole the economic effects, as seen in the overall growth in firm employment attributed to the SBIR awards, are not statistically significant, the effects are fairly large absolutely and in terms of the employment effects per million dollars of funding. Table 5.4 compares, by agency, employment gains per sampled project to the level of funding, and Table 5.5 shows the total employment gain for all of the sampled SBIR recipients in each agency. As shown in Table 5.4, for the two agencies where the mean employment gains are largest, the mean

Table 5.4 Mean SBIR-Induced "Employment Gain" by Agency (standard deviation)

Agency	*n*	Mean "employment gain"[a]	Mean "employment gain" per million $ awarded[b]
DoD	755	28.54 (89.78)	45.53 (163.61)
NIH	391	41.31 (291.33)	65.00 (378.47)
NASA	155	13.83 (67.80)	24.40 (118.45)
DOE	140	21.60 (114.91)	32.51 (182.07)
NSF	141		

NOTE: Blank = not applicable.

[a] "Mean 'employment gain'" is the average difference between actual and predicted employment in 2005 for Phase II SBIR award recipients (rather than the difference in the actual and predicted natural logarithms as given by *diff*). The firms with more than one Phase II award sampled provide multiple experiments from which to estimate the impact of an award on overall firm growth. However, those experiments produce forecasts of predicted employment in 2005 that are not independently distributed (recall that the errors in equation in the growth model were assumed to be correlated for sampled awards won by the same firm). To evaluate the significance of the average employment gains reported in this table, we need the standard error for the average forecast. It is the square root of the estimated variance of the average forecast. The average forecast is the weighted sum (with each weight being $1/n$) of the n individual forecasts. The variance of the average forecast is then a double sum over the weighted (with each weight being $1/n^2$) covariances of the forecasts. Because not all of the n forecasts are independently distributed, the standard error of the average forecast is not simply the square root of $1/n^2$ (the sum of the estimated variances for the n awards). Accounting for the covariances (when there are multiple projects from the same firm) yields a standard error for each of the average gains that is somewhat larger than the average gain itself; hence, the average gains are not statistically different from zero.

[b] "Mean 'employment gain' per million $ awarded" is defined for each sampled award as the "employment gain" for the firm winning the award divided by the amount of the Phase II award in millions of dollars.

SOURCE: Authors' calculations.

Table 5.5 SBIR-Induced "Employment Gain" by Agency, Using Just a Single Sampled Award for Each Firm

Agency	n	Mean "employment gain" (standard deviation)	Total "employment gain" for responding firms
DoD	487	22.92 (90.60)	487 × 22.92 = 11,162
NIH	290	32.40 (243.67)	290 × 32.40 = 9,396
NASA	123	12.62 (70.75)	123 × 12.62 = 1,552
DOE	110	19.10 (99.26)	110 × 19.10 = 2,101
NSF			

NOTE: Blank = not applicable. For the responding firms with multiple awards, just the first (i.e., oldest) sampled award was used in this table.
SOURCE: Authors' calculations.

firm employment gain per million dollars of funding estimated with each sampled Phase II project is 45.5 for the DoD and 65.0 for the NIH.

In Chapter 6 we take advantage of the wide range of outcomes to investigate empirically variables that are correlated with induced employment growth. In other words, we ask if there are particular firm or project characteristics that are more likely than not to be associated with SBIR projects that result in positive and greater-than-predicted employment growth.

Notes

1. There are two employment data points in the NRC database, the number of employees at the time of the Phase II award and employment in 2005 (after the Phase II research is completed).
2. This model is based on Link and Scott (2003, 2006).
3. The dependent variable is the natural logarithm of the number of employees (*empt*), adding one for the owner/manager/founder/entrepreneur at the time of the Phase II award, t years after the firm was founded. We have added one to the number of employees to include the owner/manager or the founder; especially for the younger firms the addition is of substantive importance. Obviously someone was employed at the firm and filling out the survey forms in those firms reporting zero employees. We have interviewed the owner/manager/founder/entrepreneur of SBIR firms where that one individual was the firm's president and CEO as well as the chief researcher; he or she "wore all the hats." See Link and Scott (2000) for case study examples.
4. As we stated in Chapter 1, to our knowledge our analysis is the first to use a counterfactual model to quantify the relationship between investments in innovation

through public support of R&D and employment growth in firms of any size. As such, our empirical analysis, especially our analysis in this chapter, is exploratory in nature.

5. Firm size is not a variable in the model of growth because we estimate it with observations at two times—the beginning, when all firms are the same size, and then using size at the time of the award. There is a rich literature related to testing Gibrat's Law (see Sutton [1997] for a critique of this literature). Empirically, models that test Gibrat's Law that firm growth is proportional to firm size include in the regression model base firm size, S_t, as a regressor. In our model, all firms' base size, where size is measured in terms of employment, is the parameter *a* in Equation (5.1). Thus, with the NRC data a comparable test of Gibrat's Law is not possible.
6. In the growth model estimated on preaward information, the age of the firm (*firmage*) at the time of the Phase II award is measured in years (t) (see the Glossary of Variables and Descriptive Statistics, Table A.1). When the estimated employment model is used to predict the counterfactual employment for the firm in 2005, the variables firm age and time are of course growing and the predicted annual growth rate for the firm is changing as the firm ages.
7. The number of previous Phase II awards might be associated with types of firms that are especially likely to grow rapidly given previous SBIR support. Yet, alternatively, such experience could be associated with "SBIR mills" that may have more interest in receiving multiple awards than in pursuing the commercial applications of those awards (Wessner 2000).
8. For example, Stinchcombe (1965) is often credited with the general observation that the founding history of an organization (e.g., a firm) directly influences the present structure of the organization. Baron, Hannan, and Burton's (1999, p. 529) interpretation of Stinchcombe's theory is that "founding conditions become imprinted on organizations and mold their subsequent development."
9. Link and Scott (2009, p. 269) observe, "On the one hand, experience in business is expected to be positively associated with commercialization. Yet, on the other hand, the SBIR program could be seen as looking for business entrepreneurs who have innovative ideas yet would not (because of market failures that could include the financial market's failure to support some entrepreneurs lacking business experience yet having socially desirable projects) receive sufficient R&D funding from the private sector alone. Given the SBIR support, such award recipients, vigorously pursuing and championing perhaps their first foray into business, may actually have greater success in commercializing than award winners with business backgrounds."
10. Richardson (1972, p. 888) argues, "The capabilities of an organisation may depend upon command of some material technology . . . [and] organisations will tend to specialize in activities for which their capabilities offer some comparative advantage . . . "
11. The unique characteristics of firms, and hence their unique behaviors and performances, even after controlling for the characteristics of the industries in which they operate, were first demonstrated convincingly for large samples of lines of business for manufacturing firms in Scott (1984) and Scott and Pascoe (1986).

12. Augier and Teece (2007, p. 179) observe, “Dynamic capabilities refer to the (inimitable) capacity firms have to shape, reshape, configure and reconfigure the firm’s asset base so as to respond to changing technologies and markets. Dynamic capabilities relate to the firm’s ability to proactively adapt in order to generate and exploit internal and external *firm specific competencies*, and to address the firm’s *changing environment*.”
13. See National Association of Seed and Venture Funds (2006).
14. This variable is the firm’s number of Phase II awards in the sampling pool.
15. The model is estimated using StataCorp (2007).
16. An estimate (rho) of the correlation between the error in the model of employment and the error in the selection and an estimate (sigma) of the standard deviation of the error in the employment model are provided and subsequently used with the probability density and probability of selection, as well as the estimates from the employment model, to provide predictions of employment conditional on response to the NRC survey.
17. The assumptions underlying this calculation are *sbirfnd* = 0, *otherstarts* = 0, *busfndrs* = 1, *acadfndrs* = 1, *lateeval* = 1, *south* = 1, and *prevphII* = mean value from the sample. (See the descriptive statistics for the variables, Table A.2.)
18. $empt^{A}_{05} = (empt)e^{gpost(tsa)}$.
19. The formula for the expected value of the dependent variable conditional on response to the survey is developed and given by Greene (2003, p. 784). Briefly, the conditional expectation is the inner product of the values for the explanatory variables and the estimated parameters for the employment model plus the product of three things: 1) the correlation (rho) of the error in the employment model and the error in the selection model, 2) the estimate (sigma) of the standard deviation of error in the employment model, and 3) the ratio of the probability density of response to the probability of response. Thus, we use the “ycond” option with the postestimation command “predict” for the maximum likelihood Heckman selection model in StataCorp (2007). Therefore, “ycond” calculates the expected value of the dependent variable, conditional on the dependent variable being observed (i.e., selected).

6
Factors Related to Employment Growth from SBIR Funding

In this chapter we continue our discussion of the dynamic issue of whether the Small Business Innovation Research (SBIR) program assists small entrepreneurial firms in clearing the initial hurdles they face and achieving growth beyond what would be possible if they did not have the support of the SBIR program. Specifically, we ask, "What characteristics of a firm explain the difference between its actual employment achieved with the SBIR award and the counterfactual baseline of its predicted employment in the absence of the award?" Or, stated differently, "In what types of firms does SBIR funding induce employment growth?"

Phrased either way, this is an important question for at least two reasons. First, from a policy perspective, should employment growth ever become an explicit objective of the SBIR program under a future reauthorization—or should a program similar to SBIR be initiated at some point to stimulate technology-based employment growth—our findings about firm and market characteristics that are correlated with greater-than-predicted employment growth might provide some award criteria relevant for consideration. And second, from an academic perspective, the analysis in Chapter 5 anticipates examining why some firms have grown faster than their predicted growth rate.

To answer this question, we here examine the difference, *diff*, between the natural logarithms of the actual (A) employment in 2005 and the predicted (P) employment in that year. The prediction for employment was the outcome of the counterfactual model estimated in Chapter 5. Thus, we estimate a model of SBIR-induced employment growth for each of the five agencies. Our performance model is

(6.1) $diff = f(X)$,

where the market and firm characteristics in vector X are described below.

For our exploratory analysis of cross-firm differences in actual employment relative to predicted employment, parsimonious models are estimated that, apart from including all of the commercial-type variables considered in Chapter 4, use only those variables with effects that are significantly different from zero.[1] These parsimonious models are estimated for four reasons. First, the samples for the different agencies differ markedly in the variables explaining induced employment growth. Second, many of the variables describing the firms do not have statistically significant effects. Third, the samples for which all of the explanatory variables are available are small, and reducing the number of parameters estimated allows models with more substantial degrees of freedom. Fourth, unlike the variable on retained employees studied in Chapter 4, our greater-than-predicted employment growth variable is a calculated variable rather than one coming from observed data. Therefore, the analysis that follows is quite novel and exploratory, and the initial complete specifications (from which the parsimonious specifications were selected) would confront the reader with a large number of insignificant variables. For those insignificant variables, there is no strong a priori argument to expect significance, yet they are included in the entire collection of potential explanatory variables—characterizations of the firms and their projects—from which the exploratory analysis found the significant ones.

CONSTRUCTION OF NEW SUMMARY VARIABLES

For the parsimonious models estimated in this chapter, we introduce two summary variables that are derived from the variables introduced and used in the earlier chapters.

The first of these new summary variables is the amount of outside finance (i.e., not provided by the firm itself or its principals, other than the funding provided by the SBIR program) relative to total investment in the Phase II project (*outfintoinv*).[2] Note that this constructed outside financing variable does not include the funding from colleges and universities. Funding from these sources will be controlled separately because the funding from academia is arguably more analogous to direct research support, rather than being simply generic funding for the operation of the project.

The second of the constructed variables is derived from the 11 variables describing commercial agreements.[3] The variable (*dagreement*) is a 0/1 variable that equals 1 whenever any one or more of the 11 commercial agreement variables for which we have information equals 1, and is zero otherwise. In other words, this new summary variable simply controls for the presence of some type of commercial agreement.

Table 6.1 provides the descriptive statistics for these two new summary variables for each of the five agencies. In the parsimonious models, the summary variables for outside funding and for commercial agreements are used if they have statistically significant effects. If a summary variable does not have a statistically significant effect, then variables from which the summary variable was constructed are used when significant.[4] For some agencies, a summary variable was insignificant because some of its underlying variables had significant effects of opposite signs. We did not employ this estimation strategy, which

Table 6.1 Descriptive Statistics for Constructed Variables for Outside Finance and Commercial Agreements: Mean, (Std. dev.), {range}, [*n*]

Variable	DoD	NIH	NASA	DOE	NSF
Outside	0.1489	0.0907	0.0754	0.1496	0.2181
financing	(0.2534)	(0.2233)	(0.1711)	(0.2593)	(0.2927)
(*outfintoin*)	{0–0.9869}	{0–0.9952}	{0–0.8000}	{0–0.9698}	{0–0.9796}
	[755]	[389]	[155]	[137]	[137]
Commercial	0.6708	0.7325	0.5872	0.6789	0.8649
agreements	(0.4703)	(0.4434)	(0.4946)	(0.4691)	(0.3434)
(*dagreement*)	{0/1}	{0/1}	{0/1}	{0/1}	{0/1}
	[562]	[314]	[109]	[109]	[111]

NOTE: Here, *n* is for the sample for which *diff* and the constructed variable are both available. The descriptive statistics for the variables (Table A.2) show *n* for the underlying funding variables and underlying commercial agreement variables without restriction to the subset for which *diff* is also available. In the regressions below, *n* is further restricted to the subset of those observations that not only have both *diff* and the finance and agreement variables, but also have the other explanatory variables, none of which were used in the growth model from which *diff* was derived.
SOURCE: Authors' calculations.

we argue is appropriate here, in Chapter 4 because of the fourth reason for the parsimonious models that we explained and elaborated on in the introduction to this chapter—namely, that our dependent variable in the performance model in Equation (6.1) is a constructed variable.

From the descriptive statistics in Table 6.1, NIH and NASA firms, on average, rely less on outside funding than do DoD, DOE, or NSF firms. In fact, NSF firms, on average, rely two to three times more on outside funding than do NIH and NASA firms. There is not as much difference among firms, by agency, in the presence of a commercial agreement. Near 60 percent of NASA firms are so engaged, compared to about 70 percent of DoD, NIH, and DOE firms, and to about 85 percent of NSF firms.

EXPLANATORY VARIABLES AND THE ESTIMATED EFFECTS

The estimated parsimonious performance model for each of the five agencies' samples of SBIR projects is found in Table B.7 in the Technical Appendix. The explanatory variables in the models are variables that were not used in the growth model in Chapter 5 to make the predictions that we analyze here. In the specifications of this chapter, we explore other variables determining the part of actual employment, $\ln(empt^{A}_{05})$, that is not predicted from the employment model in the previous chapter, $\ln(empt^{P}_{05})$.[5]

Below, we discuss the estimated results by category of variables.

Firm characteristics. An observed positive effect of a firm having prior spin-off companies from SBIR funding (*spinoffs*) on the trajectory of employment growth suggests that the firm has assembled a resource base capable of creating new opportunities for a commercially viable outcome from the SBIR Phase II project sampled. Those opportunities can support growth beyond what would have been predicted without the award. Empirically, this is the case for DoD and NIH projects. Yet a negative effect could, in concept, be observed. A firm with spin-off experience might choose to stay focused on developing new technologies and not exploit the potential for growing employment to support manufacturing using the Phase II project's technology. However, no estimated negative effect was observed.

Regarding the percentage of revenues taken by awarded funds (*percrevssbir*), a negative relationship, as found for the DoD sample, might signal that a firm is still largely at a precommercial R&D stage and has not yet begun growing employment to support commercial sales. Although only observed for NSF projects, the relationship could, in concept, be positive. Perhaps the R&D of those projects had resulted in a commercially viable technology for which the firm was rapidly growing employment to support the final market development effort.

SBIR research variables. DoD, NIH, and NSF projects in firms that have had previous Phase II SBIR awards related to the current award (*rltdphII*) realized employment growth greater than projected. This positive relationship could reflect a critical mass of developed technology that is complementary to the technology from the Phase II award sampled; such complementary technology might be expected to support strong commercial growth for the firm exploiting the technology developed with the sampled project. Its effect is not negative in any of the samples—although, if the firm has related technology, it is possible that the newly developed technology from the sampled project would not add appreciable commercial potential for the firm, and the impact could then be negative.

Market variables. Only in the NASA sample is there a significant relationship between the presence of a federal acquisition program (*fedacq*) and the firm's employment trajectory. In this case, the SBIR award increases the employment growth trajectory of the firm because the government provides a market for the commercial results from the SBIR project. The effect is never negative for any of the other four samples.

Intellectual property variables. When the SBIR project results in intellectual property (*patents, copyrights, trademarks, publications*), a positive effect on the difference between actual and predicted growth could be expected because such activity signals commercially valuable technology and such activity provides some protection of the commercial value of the technology. Positive effects for intellectual property variables are found in the DoD, NIH, NASA, DOE, and NSF samples. To the extent that publications—a public good—reveal commercially valuable information that others (e.g., competitors) could use, the effect could be negative. Evidence supporting that possibility is found for just the NIH sample.

Funding variables. The proportion of outside funding for the project that comes from sources other than the SBIR program (*outfintoinv*) signals that others believe in the commercial viability of the project. This variable, or one of its underlying variables, is associated with a positive impact on extraemployment growth in the DoD, NIH, and NSF samples. When the effect is negative, as for some of the underlying outside finance variables in the NSF sample, outside funding might still be a signal of commercial viability but not of the need for employment growth above the projected level.

Economic theory predicts over time a positive relationship between outside funding and greater-than-predicted growth. Åstebro (2003) argues that independent inventors, who in some respects are similar to small entrepreneurial firms, face problems in attracting outside investors because of information asymmetries. When outside investors do invest in a potential technology, it is logical to conclude that at least two hurdles have been cleared. The first hurdle is that the firm receiving the research funds was selected to be scrutinized from all possible firms, and the second hurdle is that it was selected from all those that were scrutinized. The presence of outside funding suggests that the firm's technology is promising, and thus one would expect the number of employees to be above what would have been predicted based on preaward information.[6]

The firm's own funding in the project (*own-firm-funding-to-total-investment*) signals an internal commitment and one that is revealed in employment growth, as in the NIH, NASA, and NSF samples. In contrast to the use of own-firm funding, an increase in personal funding (*personal-funds-to-total-investment*) is associated with less-than-projected employment growth in the DoD, NIH, and DOE samples and does not have a statistically significant effect on projected employment growth for the NASA and NSF samples. Perhaps increases in personal funding reflect insight on the part of the firm's principal that an internal adjustment is needed to ensure commercial success and attract outside funding.

If funding support from academia (*colleges-or-universities-funding-to-total-investment)* is a higher proportion of investment, the underlying R&D might be more basic—more toward the research and less toward the development—and either less certain to yield commercially viable results in the near term or less understood by outside investors. In any event, those projects with a greater proportion of academic investments have lower-than-projected employment growth.

Commercial agreements. *A priori*, the presence of commercial agreements regarding the technology developed with the SBIR Phase II award (*dagreement*) should be an important factor in explaining the broad effect of the award on the firm's employment growth. Such firms that are involved in manufacturing agreements might allow their partners to hire the employees. If this is the case, then the presence of such agreements would be negatively related to internal SBIR-induced employment growth and would suggest that the impact of SBIR funding on such employment growth is at least one step removed from the researching firm. A negative effect for the summary commercial agreements variable is found in the DoD and DOE samples. The negative effect for some of the individual variables underlying the summary commercial agreement variable is present for NIH, NASA, and NSF projects. In such cases, the SBIR firm may choose to focus on developing new technology, which means it may use its current resource base to develop additional technology rather than grow to meet the employment needs of manufacturing based on the technology already developed. This finding should not be interpreted to mean that the firm's SBIR project was less successful than expected. Rather, licensing the use of the technology to others may, for example, help the SBIR firm penetrate markets rapidly and increase demand for its product during a window of opportunity that can be very short-lived in markets using rapidly evolving technologies. Positive effects for some of the underlying commercial agreement variables are found in the NASA and NSF samples.

Commercial-type variables. The statistical models control for all of the commercial-type variables, even though most do not exhibit significant effects, to ensure that effects for different types of commercial activity are not restricted to being the same. Models of economic performance for firms have typically allowed the data to show different performance for different types of activity, *ceteris paribus*. In the DoD, NASA, and DOE samples, when an SBIR project plans to commercialize educational materials (*education*), employee growth is significantly less than what would be predicted in the absence of the project.

INTERPRETATION OF THE FINDINGS

Because of the exploratory nature of the performance model and empirical analysis in this chapter, we exercise caution in interpreting the results from a policy perspective. It is tempting, however, to ask if the analysis suggests a policy lever that might be pulled *prior* to an award if an implicit goal of the SBIR program is to induce employment growth within the firm. Evidence of investments in intellectual property at the end of Phase I might therefore indicate a greater likelihood that the Phase II research will result in a positive employment trajectory for the firm, and might therefore be taken into consideration when selecting Phase II award recipients. However, caution is necessary, because the significant positive relation between intellectual property and the SBIR employment impact on the award-recipient firm quite likely entails intellectual property developed during the Phase II project. Also, any such preaward policy might thwart the possibility of employment growth in other firms with which the SBIR firm has an after–Phase II commercial agreement. We are doubtful whether such a possibility could be predicted prior to a Phase II award.

Notes

1. For each agency, an initial regression included the full set of variables. Variables with statistically insignificant coefficients (excepting the commercial-type variables) were then dropped to leave the parsimonious specification.
2. The definition, in terms of previously defined variables, is as follows: *outfintoinv* = *non-sbir-federal-to-total-investment* + *U.S.-private-venture-capital-to-total-investment* + *foreign-private-investment-to-total-investment* + *other-private-equity-investment-to-total-investment* + *other-domestic-private-firm-investment-*

to-total-investment + *state-or-local-government-funding-to-total-investment*. The variables used to construct this summary variable are defined in the Glossary of Variables and Descriptive Statistics, Table A.1.

3. The 11 commercial agreement variables are defined in the Glossary of Variables and Descriptive Statistics, Table A.1.
4. The regression with all of the variables used the summary variable. If it was not significant, the initial full regression was estimated using not the summary variable but its underlying variables. Then, insignificant variables were dropped to reach the parsimonious specification.
5. The variables used in the growth model are already present in the formulation of $diff = [\ln(empt^{A}_{05}) - \ln(empt^{P}_{05})]$ because they determine $\ln(empt^{P}_{05})$.
6. Related to this, Zeira (1987, pp. 204–205) argues that such outside funding reduces what he calls "structural uncertainty," a situation in which the firm does not fully know its own profit function, which relates profits to capital stock. "The exact level of profits at larger amounts of capital can be discovered only by actually increasing the quantity of capital, that is, by 'getting there.' Investment under structural uncertainty creates an interaction between the accumulation of capital and the accumulation of information . . . In the face of structural uncertainty, investment becomes a process of learning the profit function and of searching for the optimal amount of capital."

7
The Exploitation of SBIR-Induced Intellectual Capital and Employment Growth

Firms pursue profit maximization by relying on numerous strategies. Examples of such strategies include, but are certainly not limited to, the development of a new technology with the expectation of selling or licensing it to other firms or even an expectation of being acquired by or merging with other firms. Alternatively, firms can pursue further research through R&D agreements. And, firms can attempt to realize more fully the profit potential of their technology by entering into joint ventures (all types), manufacturing/marketing/distribution agreements, or customer agreements. Any or all of these strategies should be viewed as a rational effort on the part of the firm to maximize profit.

The empirical results in Chapter 6 show that the commercial agreements reflecting the foregoing strategies are often negatively associated with greater-than-projected employment growth within the firm. The Small Business Innovation Research (SBIR) firm develops the technology during the Phase II project but, for example, may sell the rights to the technology to another domestic firm. The employment impact of the SBIR award will then be felt elsewhere in the economy where another firm has employment growth to support production using the technology developed in the SBIR project. Thus, relevant to the descriptive analysis in this chapter, which is motivated by the commercial agreement variables in the analyses in Chapters 4 and 6, is the question of whether award-recipient firms pursue such strategies using commercial agreements with other U.S. firms or investors, or even with foreign firms or investors. In other words, the question asked is, "Is the employment impact of the SBIR program confined to U.S. firms, or is it being realized to any significant extent among foreign firms?" Stated still another way, "Is there evidence that the technologies developed with funds from U.S. taxpayers allocated through the SBIR program are being exploited, through commercial agreements, by foreign firms?"[1]

INTERPRETATION OF THE DESCRIPTIVE STATISTICS

The descriptive statistics presented in this chapter should be viewed as only a first step toward addressing these important issues. As shown in Table 7.1, there are similarities and differences across agencies in the extent to which alternative commercial agreements are in place among U.S. and foreign firms or investors. Regarding agreements with other U.S. firms or investors, the primary strategies being finalized or negotiated, across all agencies, are licensing agreements, marketing/distribution agreements, R&D agreements, and customer alliances (but not necessarily in that order).

While the NRC survey data are insufficient to quantify the magnitude of any indirect employment effects, much less competitiveness effects, from the existence of these activities, our findings from Chapter 6 do suggest that SBIR funding may leverage employment gains when agreements are undertaken with other firms or investors. Complementing these findings, there is evidence from case studies of research joint-venture and R&D alliances that the parties cooperate on projects along a particular research path, but at the same time they pursue a related but independent research path as a diversification strategy (Link 2006). To the extent that this happens among SBIR firms and their partners, then direct and indirect employment effects might be possible.

Four generalizations can be inferred from the descriptive statistics in Table 7.1. One, commercial agreements are more common among U.S. firms and investors than among foreign firms and investors. There are only two instances where finalized agreements with foreign firms or investors are greater than with U.S. firms or investors. Among the firms in the NASA sample, 8 percent had formalized a licensing agreement with foreign firms or investors, compared to 6.2 percent with domestic firms or investors. And, 1.7 percent of the firms in the NSF sample had formalized a merger with foreign firms or investors, compared to 0.0 percent with U.S. counterparts.

Two, among U.S. firms and investors, mergers and sale of the company are the least-used strategies across all of the funding agencies. This compares to licensing agreements being the most-used strategy by SBIR firms funded by the DoD, NIH, DOE, and NSF. Following licensing agreements by firms and investors in these four agencies are

Table 7.1 Agreements with Other Firms or Investors: Percentage of the Random Sample of Phase II Projects Answering the Question, "As a result of the technology developed during this project, which of the following describes your firm's activities with other firms and investors? (Select all that apply.)"

DoD (n = 594)	U.S. firms/investors		Foreign firms/investors	
Agreement	Finalized agreements	Ongoing negotiations	Finalized agreements	Ongoing negotiations
Licensing agreement(s)	15.5	16.5	2.9	5.2
Sale of company	1.2	4.9	0.3	1.0
Partial sale of company	0.8	4.2	0.0	1.5
Sale of technology rights	4.4	9.9	1.0	2.9
Company merger	0.3	3.2	0.2	0.7
Joint venture agreement(s)	3.7	8.1	1.2	2.2
Marketing/ distribution agreement(s)	10.8	8.6	4.5	4.2
Manufacturing agreement(s)	3.4	8.8	2.7	2.4
R&D agreement(s)	13.8	13.8	2.5	3.4
Customer alliance(s)	13.1	14.5	4.2	3.0
Other	1.9	2.2	0.3	0.8

(continued)

Table 7.1 (continued)

NIH ($n = 338$)	U.S. firms/investors		Foreign firms/investors	
Agreement	Finalized agreements	Ongoing negotiations	Finalized agreements	Ongoing negotiations
Licensing agreement(s)	19.2	16.0	8.9	6.2
Sale of company	0.9	3.6	0.3	1.5
Partial sale of company	2.1	4.4	0.0	0.6
Sale of technology rights	5.6	7.4	0.6	0.9
Company merger	0.3	3.0	0.0	0.6
Joint venture agreement(s)	3.0	8.9	0.9	2.7
Marketing/distribution agreement(s)	21.3	10.4	12.4	6.5
Manufacturing agreement(s)	7.1	3.8	2.4	2.1
R&D agreement(s)	14.8	10.7	4.1	3.0
Customer alliance(s)	8.3	10.1	2.7	0.9
Other	2.1	2.1	0.3	0.9

Table 7.1 (continued)

NASA (n = 112)	U.S. firms/investors		Foreign firms/investors	
Agreement	Finalized agreements	Ongoing negotiations	Finalized agreements	Ongoing negotiations
Licensing agreement(s)	6.2	11.6	8.0	5.4
Sale of company	0.9	2.7	0.0	0.0
Partial sale of company	1.8	0.9	0.0	0.0
Sale of technology rights	0.9	7.1	0.9	2.7
Company merger	0.0	0.0	0.0	0.0
Joint venture agreement(s)	0.9	4.5	0.0	1.8
Marketing/distribution agreement(s)	7.1	5.4	6.2	1.8
Manufacturing agreement(s)	1.8	6.2	1.8	0.0
R&D agreement(s)	15.2	9.8	3.6	0.9
Customer alliance(s)	10.7	8.9	6.2	0.0
Other	2.7	3.6	0.9	0.9

(continued)

Table 7.1 (continued)

DOE (n = 114)	U.S. firms/investors		Foreign firms/investors	
Agreement	Finalized agreements	Ongoing negotiations	Finalized agreements	Ongoing negotiations
Licensing agreement(s)	15.8	15.8	5.3	8.8
Sale of company	0.9	0.9	0.9	0.9
Partial sale of company	2.6	1.8	0.0	1.8
Sale of technology rights	5.3	6.1	0.9	3.5
Company merger	0.0	0.9	0.0	0.9
Joint venture agreement(s)	2.6	7.0	0.0	2.6
Marketing/distribution agreement(s)	9.6	7.0	8.8	2.6
Manufacturing agreement(s)	6.1	6.1	0.9	4.4
R&D agreement(s)	7.9	10.5	1.8	6.1
Customer alliance(s)	7.9	13.2	4.4	5.3
Other	2.6	0.0	0.9	0.0

Table 7.1 (continued)

NSF (n = 121)	U.S. firms/investors		Foreign firms/investors	
Agreement	Finalized agreements	Ongoing negotiations	Finalized agreements	Ongoing negotiations
Licensing agreement(s)	19.8	21.5	9.9	6.6
Sale of company	2.5	3.3	0.0	0.8
Partial sale of company	2.5	5.8	1.7	1.7
Sale of technology rights	5.0	15.7	4.1	3.3
Company merger	0.0	4.1	1.7	0.8
Joint venture agreement(s)	3.3	9.9	0.8	2.5
Marketing/ distribution agreement(s)	15.7	12.4	8.3	2.5
Manufacturing agreement(s)	8.3	9.9	3.3	2.5
R&D agreement(s)	17.4	17.4	5.0	6.6
Customer alliance(s)	11.6	18.2	3.3	4.1
Other	1.7	2.5	0.8	0.0

NOTE: The overall percentage of projects that have agreement activities is less than would be the case if the percentages of projects were additive across the types of categories. They are not additive. Thus, for example, for the DoD, it will not in general be the case that 16 percent of the Phase II projects have finalized U.S. licensing agreements while a completely different set of 14 percent have finalized U.S. R&D agreements and yet another distinct 13 percent have finalized U.S. customer alliances. To determine the percentage of the random sample of Phase II projects with U.S. agreements, with foreign agreements, and with both U.S. and foreign agreements, we constructed the qualitative variables defined in Table 7.2. Those variables are defined for all agencies.

SOURCE: Authors' tabulations.

R&D agreements—R&D agreements rank first among NASA firms. In addition, licensing agreements with U.S. firms and investors are at least twice as common as those with foreign firms and investors, with the exception of the NASA sample.

Three, licensing agreements and marketing/distribution agreements are the more common strategies with foreign firms or investors.

And four, R&D agreements are significantly more common with U.S. firms and investors than with foreign firms and investors. Perhaps—and this should be viewed as a tentative hypothesis—it is easier to monitor information leakages that are inevitable during collaborative R&D when the partner firms or investors are geographically closer and within the same boundaries of intellectual property protection—that is, with a U.S. partner rather than a foreign partner.

As shown in Table 7.2, U.S.-only agreements are more prevalent than U.S. and foreign firm or investor agreements. Foreign-only agreements are the least prevalent. Thus, we interpret the data in Tables 7.1 and 7.2 as preliminary evidence to conclude that U.S. SBIR funds for Phase II projects and the associated technologies and employment growth associated with those projects are not to a pronounced extent benefiting foreign firms through agreements with SBIR firms or investors. In that sense, there is no evidence that the technologies developed with funds from U.S. taxpayers are, to any significant extent, being exploited by foreign firms through agreements with SBIR firms. If there were such evidence of a substantial flow of SBIR technology to foreign firms, one could reasonably ask whether the U.S. SBIR firms captured a substantial part of the benefits from the applications of the SBIR technology by the foreign firms.[2] But that question need not be asked.

Finally, in Table 7.3, we examine the use of alternative strategies by observing the level of sales from the specific technology developed during the Phase II project. Regardless of funding agency, the mean sales for those projects with no agreements—neither U.S. nor foreign agreements—are less than for projects with agreements. And, mean sales are greater when there are agreements with foreign firms or investors than when there are only U.S. agreements. This finding, albeit based only on descriptive evidence—and in some cases descriptive evidence based only on a handful of observations—suggests that when such technology transfer does occur through commercial agreements among SBIR firms and foreign firms and investors, it is with projects that are relatively

Table 7.2 Descriptive Statistics for the Presence of Agreements

DoD	*n*	Mean	Std. dev.	Min.	Max.
U.S. agreement(s) only	594	0.478	0.500	0	1
Foreign agreement(s) only	594	0.0522	0.223	0	1
U.S. and foreign agreement(s)	594	0.143	0.350	0	1
NIH	***n***	**Mean**	**Std. dev.**	**Min.**	**Max.**
U.S. agreement(s) only	338	0.441	0.497	0	1
Foreign agreement(s) only	338	0.0562	0.231	0	1
U.S. and foreign agreement(s)	338	0.237	0.426	0	1
NASA	***n***	**Mean**	**Std. dev.**	**Min.**	**Max.**
U.S. agreement(s) only	112	0.366	0.484	0	1
Foreign agreement(s) only	112	0.0446	0.207	0	1
U.S. and foreign agreement(s)	112	0.179	0.385	0	1
DOE	***n***	**Mean**	**Std. dev.**	**Min.**	**Max.**
U.S. agreement(s) only	114	0.395	0.491	0	1
Foreign agreement(s) only	114	0.070	0.257	0	1
U.S. and foreign agreement(s)	114	0.228	0.421	0	1
NSF	***n***	**Mean**	**Std. dev.**	**Min.**	**Max.**
U.S. agreement(s) only	121	0.521	0.502	0	1
Foreign agreement(s) only	121	0.083	0.276	0	1
U.S. and foreign agreement(s)	121	0.240	0.429	0	1

NOTE: U.S. agreement(s) only: 0/1, with 1 indicating agreement(s)—finalized or ongoing negotiations—with U.S. companies or investors only. Foreign agreement(s) only: 0/1, with 1 indicating agreement(s)—finalized or ongoing negotiations—with foreign companies or investors only. U.S. and foreign agreement(s): 0/1, with 1 indicating both U.S. and foreign agreements.

SOURCE: Authors' calculations.

Table 7.3 Descriptive Statistics for Sales Categories with and without U.S. and Foreign Agreements

DoD	*n*	Mean	Std. dev.	Min.	Max.
Sales for the Phase II projects providing information about agreements	594	1,763,832	9,659,895	0	201,000,000
Sales with no agreements	194	972,675	3,941,742	0	49,300,000
Sales with U.S. agreement(s) only	284	1,116,917	3,264,621	0	35,000,000
Sales with foreign agreement(s) only	31	1,992,457	6,504,660	0	35,000,000
Sales with U.S. and foreign agreements	85	5,647,607	2.35×10^7	0	201,000,000
NIH	***n***	**Mean**	**Std. dev.**	**Min.**	**Max.**
Sales for the Phase II projects providing information about agreements	338	2,003,032	1.01×10^7	0	100,000,000
Sales with no agreements	90	320,273	800,612	0	5,000,000
Sales with U.S. agreement(s) only	149	1,061,330	4,849,109	0	40,000,000
Sales with foreign agreement(s) only	19	6,787,637	1.94×10^7	0	81,800,000
Sales with U.S. and foreign agreements	80	4,513,711	1.70×10^7	0	100,000,000
NASA	***n***	**Mean**	**Std. dev.**	**Min.**	**Max.**
Sales for the Phase II projects providing information about agreements	112	917,510	2,313,071	0	15,100,000
Sales with no agreements	46	430,520	853,384	0	4,004,011
Sales with U.S. agreement(s) only	41	938,914	2,375,296	0	13,500,000
Sales with foreign agreement(s) only	5	2,616,600	4,518,742	0	10,600,000
Sales with U.S. and foreign agreements	20	1,568,936	3,402,045	0	15,100,000

Table 7.3 (continued)

DOE	*n*	Mean	Std. dev.	Min.	Max.
Sales for the Phase II projects providing information about agreements	114	1,024,187	3,023,294	0	23,400,000
Sales with no agreements	35	286,348	715,591	0	3,500,000
Sales with U.S. agreement(s) only	45	1,376,243	4,218,549	0	23,400,000
Sales with foreign agreement(s) only	8	2,314,325	4,464,564	0	12,500,000
Sales with U.S. and foreign agreements	26	1,011,140	1,482,369	0	5,500,000
NSF	***n***	**Mean**	**Std. dev.**	**Min.**	**Max.**
Sales for the Phase II projects providing information about agreements	121	2,431,395	1.85×10^7	0	203,000,000
Sales with no agreements	19	210,230	438,816	0	1,466,617
Sales with U.S. agreement(s) only	63	3,802,841	2.56×10^7	0	203,000,000
Sales with foreign agreement(s) only	10	463,741	392,028	0	1,000,000
Sales with U.S. and foreign agreements	29	1,585,795	3,478,123	0	17,900,000

SOURCE: Authors' calculations.

more successful, as measured in terms of cumulative sales. Thus, this finding supports the belief that, for those cases, the SBIR firms are reaping substantial benefits from the technologies developed with the funds from U.S. taxpayers.

Notes

1. We are defining *exploitation* to mean that the resulting U.S.-funded technology is not being entirely used within the United States and that the benefits from the SBIR R&D are in part captured by other countries.
2. Of course, there is the possibility of spillovers of the technologies to foreign firms without any payments to the innovating SBIR firms—a possibility the NRC database does not allow us to assess.

8
Conclusions

The goals of the Small Business Innovation Research (SBIR) program, set out by Congress in the Small Business Innovation Development Act of 1982, did not include the growth of employment. In Chapter 2, we reviewed the legislated goals for the program, both in the 1982 Act and in subsequent reauthorizations. The goals include the stimulation of technological innovation, the use of small businesses to meet federal R&D needs, the encouragement of participation in technological innovation by disadvantaged persons, minorities, and women, and the increase in private sector commercialization of innovations derived from federal R&D. Although employment growth could of course be a concomitant of successful attainment of the legislated goals for the SBIR program, such growth has never been a stated objective of the SBIR program.

In this book, we analyzed, albeit in an exploratory manner within an original empirical framework, a rich and unique set of data collected by the NRC to investigate whether such public support of innovation has in fact stimulated employment growth for the firms that have received SBIR awards. Although SBIR projects are commercially successful roughly half of the time (Link and Scott 2010), possibly in many cases the employment growth that would accompany the sales from successful projects may not be observed in the firms performing those projects, but instead will occur elsewhere in the economy. Conceivably, the employment growth to support sales from technologies developed with SBIR funds will be realized in other firms that purchase or license the technology or have manufacturing agreements with the firm that performed the SBIR program research. The NRC data do not allow us to document the employment growth at firms other than the award recipients, but we are able to ask if employment growth for the award recipients is less when the recipients have made commercial agreements allowing other firms to use the technologies created by the SBIR awards.

Chapter 4 addressed the question of whether the technology created in an SBIR Phase II project has itself typically added substantially to the employment in the firm conducting the project. The evidence shows that typically the answer is no—on average the SBIR projects add only a small number of employees. The average number of employees retained because of the technology developed in SBIR projects is only one or two for the random samples of the SBIR projects for the five agencies—the DoD, NIH, NASA, DOE, and NSF. However, the range for employees retained because of the technology developed in the Phase II project is wide. Exploiting that range of outcomes, the data reveal that for many types of commercial agreements (regarding the Phase II technology) between the SBIR firm and other firms or investors, the number of employees retained is less, *ceteris paribus*, supporting the belief that the jobs created when the SBIR Phase II technology is used in production actually occur at other firms that have entered into agreements to use the technology.

Chapters 5 and 6 turned from the employment effects directly associated with the use of the SBIR project's technology to the overall trajectory of the SBIR firm's employment in order to address the dynamic question of whether the SBIR award helps the recipient firm overcome the initial hurdles that small, entrepreneurial firms face, thus facilitating a more permanent and possibly longer-term employment growth. The previous literature on entrepreneurial firms has been unable to address the question because of the lack of a comparison group to which the growth of firms receiving SBIR awards could be compared. We developed in Chapter 5 a model of the growth of small entrepreneurial firms, and with this model we estimated the growth for each of the SBIR award recipients using only information about each firm prior to when it received the SBIR Phase II award in the sample. That model is then used to predict the award recipient's ultimate employment if it had not received the Phase II award. That predicted employment absent the award then provides a baseline for comparison to address the more dynamic question.

The answer, as seen in Chapter 5, is that for four of the five agencies' samples of projects, there are substantial average employment gains that are due to the SBIR award. However, these average gains, despite being substantial in absolute amount, are not statistically significant. Nonetheless, there are a number of individual cases where the actual employment for the firm exceeds the predicted employment by a statistically significant amount. Moreover, there is a wide range in the outcomes for the employment gain attributable to the SBIR award. Chapter 6 exploited that range of outcomes to demonstrate that there are quite likely to be substantial employment gains from the SBIR program beyond those gains that we can observe using the NRC data. We expect those substantial employment gains are at other firms for two reasons: 1) under commercial agreements they use the technologies developed by the SBIR award recipients, and 2) there are many cases where the presence of commercial agreements is associated with less employment growth attributable to the SBIR projects at the firms receiving SBIR awards.

Understanding the determinants of the range in outcomes for retainees or for overall employment growth suggests some policy implications. Chapter 4's model of retainees provides large-sample econometric support for Nelson's (1982) important observations—developed from case studies—about the success of government funding of research. Other things being the same, public funding of privately performed research is more likely to lead to commercialized results (and employment gains) when the government has a direct need for the products, processes, or services developed by the research projects. In such cases, the government can provide the research direction and guidance of a potential customer for the output from a research project, and it also can provide a ready market for that output. The public funding of research is therefore more likely to stimulate employment. Chapter 6's examination of the range in the growth of overall SBIR-recipient firm employment that is attributable to the SBIR award also shows that intellectual property is an indicator of significant, in a statistical sense, employment growth. As discussed there, this finding could potentially have policy implications if investments in intellectual property by the end of Phase I are used as a criterion for Phase II financial support. However, as explained in Chapter 6, such a criterion must be applied cautiously to avoid penalizing worthwhile projects that would be expected to develop intellectual property during the Phase II portion of the project.

In sum, the impact of SBIR funding on firm employment specifically associated with the SBIR project is typically small—on average, the numbers of project-specific retainees have been small. Also, the impact on the firm's overall employment growth attributed in our counterfactual experiment to SBIR funding is small in the sense that the *average* effect on employment is not significantly different from zero.

For some individual observations of Phase II SBIR awards, the impact on retainees is large and significant. Also there are cases of large and significant impacts of a Phase II SBIR award on the overall employment growth of the firm. For both the actual number of retainees and the overall employment growth in the counterfactual experiment, there is a wide range of outcomes across the individual SBIR projects for each agency. Moreover, what impact there is varies systematically across agencies based on distinct project and firm characteristics.

Just as Lerner (2010) finds that some public programs to stimulate high-potential business ventures have failed while others have succeeded, within the SBIR program which we study by looking at many of its awards we find a range of outcomes, from unsuccessful to successful, for the program's effects on employment. Moreover, just as Lerner (2010) finds that some practices favor success while others are associated with failure, we show that a set of variables describing the circumstances of the awards can significantly explain the differences in the outcomes.[1]

Chapter 7 built on the empirically based discussion in early chapters about commercial agreements. Descriptively, it appears that often the employment impact of SBIR funds is seen not in the SBIR award-recipient firm, but possibly in other firms because of commercial agreements allowing these other firms to use the technologies generated with SBIR Phase II awards. Moreover, the evidence suggests that the employment impact at times accrues to foreign firms or foreign investors through such commercial agreements. However, we observe only that, in the presence of such agreements, employment is less than it is at the SBIR award-recipient firms that do not have such agreements but are otherwise the same. It is then reasonable to assume that the firms using the SBIR technologies under commercial agreements will experience an employment impact. This is just a reasonable inference, though, because we do not directly observe the employment impact on the firms that reach agreements with the SBIR firms.

In evaluating the possibility that the SBIR awards, through commercial agreements, stimulate employment growth at other firms, recall first that employment growth is not a stated objective of the SBIR program, but commercialization is. Although, as we have shown, the employment impact, measured as project-specific retainees or overall growth in the SBIR firms, is on average small or not statistically significant, the negative impact for some commercial agreement variables (such as the sale of technology rights) supports the idea that an important part of the employment impact of the SBIR program occurs elsewhere in the national and international economies. When the recipient of an SBIR award creates new technology and sells the rights, it certainly meets the SBIR program's goal of stimulating innovation and increasing commercialization, and the employment effect—which, again, is not a stated goal of the SBIR program—will be observed at other firms.[2]

We have emphasized that although the average employment gains reported in Chapter 5 and analyzed in Chapter 6 are substantial in absolute amount, they are not statistically significant. However, in conclusion, we observe that on average, especially with respect to the age of the SBIR Phase II project that is sampled and studied, these are young firms. In many cases, enough time may not yet have passed for a strong surge in employment to have been caused by the SBIR Phase II projects that we have observed. Of course, if we had more time to observe these younger firms, they might not survive to be observed! Yet, through commercial agreements such as the sale of technology rights to other firms, the accomplishments of these young firms will live on and have employment effects that show up elsewhere in the U.S. economy and in foreign markets.

Notes

1. Importantly, as we have emphasized, generating retainees or overall employment growth of the firm is not a stated goal of the SBIR program, so an award that does not result in retainees or does not stimulate overall growth has not thereby failed to meet the SBIR program's goals.
2. Such spillover effects are consistent with the findings from case studies of SBIR awards (Link and Scott 2000).

Appendix A
Glossary of Variables and Descriptive Statistics

Descriptive statistics for the variables defined here in Table A.1 are found in Table A.2.

Table A.1 Glossary of Variables for Tables in the Book

Variable category	Definition
Firm characteristics	
t	Time from the founding of the firm until the Phase II award (i.e., the span of preaward information)
firmage	Firm's age at the time of the Phase II award, measured by *t*
firmrevenue	Firm's total revenue
empt	The number of employees (plus 1 for the owner/manager/founder/entrepreneur) *t* years after the firm was founded
emptA05	Actual number of employees in 2005
diff	Difference between the natural logarithms of the actual (A) employment in 2005 and predicted (P) employment in that year; the prediction for employment is the outcome of the counterfactual model estimated in Chapter 5
tsa	Amount of time since the Phase II award
hires	Current number of employees who were hired as a result of the technology developed during the Phase II project
retainees	Current number of employees who were retained as a result of the technology developed during the Phase II project
sbirfnd [a]	0/1, with 1 indicating the firm was founded at least in part because of the SBIR program
otherstarts	The number of other firms started by the firm's founder or founders

(continued)

Table A.1 (continued)

Variable category	Definition
busfndrs	Number of the firm's founders with business backgrounds
acadfndrs	Number of the firm's founders with academic backgrounds
phIItsvy	Firm's number of Phase II awards that were among the population of 1992–2001 projects for sampling in the survey
numsvyd	Number of the firm's Phase II projects surveyed
percrevssbir	Percentage of firm's revenues, during its last fiscal year, from federal SBIR and/or STTR funding (Phase I and/or Phase II)
spinoffs	Number of spin-off companies that the firm has established as a result of the SBIR program
SBIR research variables	
awardamt	Dollar amount of the Phase II award
prjage	Age of the Phase II project (= *tsa*)
univpartcptn	0/1, with 1 indicating that in executing the Phase II award, there was involvement by universities' faculty, graduate students, or university-developed technologies
sbir-r&d-to-total-r&d	Percentage of firm's total R&D effort (man-hours of scientists and engineers) devoted to SBIR activities during the most recent fiscal year
sbirpatents	Patents that have resulted, at least in part, from the firm's SBIR and/or STTR awards
rltdphI	Number of the firm's Phase I awards that are related to the Phase II project as of the time of that project
rltdphII	Number of the firm's Phase II awards that are related to the Phase II project as of the time of that project
prevphII	Number of the firm's previous Phase II awards
Market variables	
fedacq	0/1, with 1 indicating that a federal system or acquisition program is using the technology from the Phase II project

Table A.1 (continued)

Variable category	Definition
lateeval [b]	0/1, with 1 indicating the firm's extant policy of late evaluation of the commercial potential for projects
Commercial-type variables[c]	
nocom	0/1, with 1 indicating no planned commercial use for project's results
software	0/1, with 1 indicating the project planned to commercialize software
hardware	0/1, with 1 indicating the project planned to commercialize hardware
process	0/1, with 1 indicating the project planned to commercialize process technology
service	0/1, with 1 indicating the project planned to commercialize a service
drug	0/1, with 1 indicating the project planned to commercialize a drug
biologic	0/1, with 1 indicating the project planned to commercialize a biologic
research	0/1, with 1 indicating the project planned to commercialize a research tool
education	0/1, with 1 indicating the project planned to commercialize educational materials
other	0/1, with 1 indicating the project planned commercialization not covered in the other categories
Geographic variables[d]	
northeast	0/1, with 1 indicating the firm is in the Northeast
south	0/1, with 1 indicating the firm is in the South
midwest	0/1, with 1 indicating the firm is in the Midwest
west	0/1, with 1 indicating the firm is in the West
Intellectual property variables	
patents	Patents applied for resulting from the technology developed as a result of the Phase II project

(continued)

Table A.1 (continued)

Variable category	Definition
copyrights	Copyrights received for the technology from the Phase II project
trademarks	Trademarks received for the technology from the Phase II project
publications	Scientific publications resulting from the Phase II project
Funding variables	
non-sbir-federal-to-total-investment	Ratio of additional developmental funding during the Phase II project from non-SBIR federal funds to total investment (including the SBIR Phase II award) in the Phase II project
U.S.-private-venture-capital-to-total-investment	Ratio of additional developmental funding during the Phase II project from U.S. private venture capital investment to total investment (including the SBIR Phase II award) in the Phase II project
foreign-private-investment-to-total-investment	Ratio of additional developmental funding during the Phase II project from foreign private investment to total investment (including the SBIR Phase II award) in the Phase II project
other-private-equity-investment-to-total-investment	Ratio of additional developmental funding during the Phase II project from other private equity investment to total investment (including the SBIR Phase II award) in the Phase II project
other-domestic-private-firm-investment-to-total-investment	Ratio of additional developmental funding during the Phase II project from other domestic private firm investment to total investment (including the SBIR Phase II award) in the Phase II project
state-or-local-government-funding-to-total-investment	Ratio of additional developmental funding during the Phase II project from state or local government funding of investment to total investment (including the SBIR Phase II award) in the Phase II project
colleges-or-universities-funding-to-total-investment	Ratio of additional developmental funding during the Phase II project from colleges or universities funding of investment to total investment (including the SBIR Phase II award) in the Phase II project

Table A.1 (continued)

Variable category	Definition
own-firm-funding-to-total-investment	Ratio of additional developmental funding during the Phase II project from the SBIR firm's (including borrowed funds) own investment to total investment (including the SBIR Phase II award) in the Phase II project
personal-funds-to-total-investment	Ratio of additional developmental funding during the Phase II project from the principals' investment of personal funds to total investment (including the SBIR Phase II award) in the Phase II project
Commercial agreement variables	
licensing agreement(s)	0/1, with 1 indicating that as a result of the technology developed during the Phase II project, the company had one or more licensing agreements finalized or in ongoing negotiations with US companies/investors or with foreign companies/investors
sale of company	0/1, with 1 indicating that as a result of the technology developed during the Phase II project, the company had the sale of the company finalized or in ongoing negotiations with US companies/investors or with foreign companies/investors
partial sale of company	0/1, with 1 indicating that as a result of the technology developed during the Phase II project, the company had the partial sale of the company finalized or in ongoing negotiations with U.S. companies/investors or with foreign companies/investors
sale of technology rights	0/1, with 1 indicating that as a result of the technology developed during the Phase II project, the company had one or more sales of technology rights agreements finalized or in ongoing negotiations with U.S. companies/investors or with foreign companies/investors

(continued)

Table A.1 (continued)

Variable category	Definition
company merger	0/1, with 1 indicating that as a result of the technology developed during the Phase II project, the company had a merger finalized or in ongoing negotiations with U.S. companies/investors or with foreign companies/investors
joint venture agreement(s)	0/1, with 1 indicating that as a result of the technology developed during the Phase II project, the company had one or more joint venture agreements finalized or in ongoing negotiations with U.S. companies/investors or with foreign companies/investors
marketing/distribution agreement(s)	0/1, with 1 indicating that as a result of the technology developed during the Phase II project, the company had one or more marketing/distribution agreements finalized or in ongoing negotiations with U.S. companies/investors or with foreign companies/investors
manufacturing agreement(s)	0/1, with 1 indicating that as a result of the technology developed during the Phase II project, the company had one or more manufacturing agreements finalized or in ongoing negotiations with U.S. companies/investors or with foreign companies/investors
r&d agreement(s)	0/1, with 1 indicating that as a result of the technology developed during the Phase II project, the company had one or more R&D agreements finalized or in ongoing negotiations with U.S. companies/investors or with foreign companies/investors
customer alliance(s)	0/1, with 1 indicating that as a result of the technology developed during the Phase II project, the company had one or more customer alliances finalized or in ongoing negotiations with U.S. companies/investors or with foreign companies/investors

Table A.1 (continued)

Variable category	Definition
other	0/1, with 1 indicating that as a result of the technology developed during the Phase II project, the company had one or more other agreements, not itemized above, finalized or in ongoing negotiations with U.S. companies/ investors or with foreign companies/investors

[a] The survey asked whether the firm was founded because of the SBIR program. The variable *sbirfnd* = 1 if the respondent said "Yes" or "Yes, in part."

[b] The survey asked when, as a matter of company policy, the firm would do an evaluation of the potential commercial market for the product, process, or service expected from an SBIR project. The variable *lateeval* = 1 if this evaluation would be done during or after the Phase II award, and 0 if it would be done prior to submitting the Phase II project.

[c] Projects for which the respondent did not provide a particular qualitative assessment of commercialization type are subsumed in the intercept term in the estimated models.

[d] Projects in the West are subsumed in the intercept term for estimated models. The geographic location of each firm was determined by U.S. Census Bureau (2010) classifications.

SOURCE: Authors' compilation.

Table A.2 Descriptive Statistics for the Variables—Mean, (Std. dev.), {Range}

Variable category	DoD (n = 755)[a]	NIH (n = 391)[a]	NASA (n = 155)[a]	DOE (n = 140)[a]	NSF (n = 141)[a]
Firm characteristics					
firmage	11.61 (10.74) {0–98}	9.32 (9.13) {0–100}	12.81 (10.83) {1–50}	11.74 (11.25) {0–97}	10.67 (12.84) {0–94}
firmrevenue	1.41×10^7 (2.66×10^7) {50,000–1.25×10^8} n = 754	8,386,375 (1.94×10^7) {50,000–1.25×10^8) n = 389	1.64×10^7 (2.93×10^7) {50,000–1.25×10^8}	1.12×10^7 (2.18 × x 10^7) {50,000–1.25×10^8} n = 139	7,061,429 (1.46×10^7) {50,000–6.00×10^7} n = 140
empt	35.46 (61.51) {1–326}	20.50 (38.98) {1–301}	45.66 (78.47) {1–376}	33.14 (60.11) {1–451}	23.54 (37.03) {1–201}
$empt^{A}_{05}$	59.93 (100.51) {1–1,001}	60.41 (293.38) {1–4,001}	62.54 (94.83) {1–521}	55.71 (118.51) {1–851}	40.99 (85.43) {1–751}
diff	0.1359 (1.172) {–5.940–4.244}	0.1201 (1.338) {–4.158–4.942}	0.02067 (1.089) {–3.723–4.509}	−0.1098 (1.218) {–4.008–3.193}	−0.3915 (1.261) {–3.935–1.938}
tsa	7.464 (2.810) {4–13}	7.233 (2.586) {4–13}	8.084 (2.923) {4–13}	7.836 (2.732) {4–13}	6.894 (2.466) {4–13}

hires	2.252	2.801	1.084	1.386	1.468
	(7.920)	(10.99)	(1.724)	(2.521)	(2.079)
	{0–150)	{0–149}	{0–10}	{0–20}	{0–12}
retainees	2.000	2.335	1.368	1.379	1.957
	(5.281)	(8.626)	(2.996)	(2.012)	(3.077)
	{0–73}	{0–149}	{0–22}	{0–15}	{0–15}
sbirfnd	0.2159	0.2583	0.2065	0.2500	0.2057
	(0.4117)	(0.4383)	(0.4061)	(0.4346)	(0.4056)
	{0/1}	{0/1}	{0/1}	{0/1}	{0/1}
otherstarts	1.301	1.194	1.787	1.443	1.631
	(2.322)	(1.930)	(3.062)	(1.957)	(2.579)
	{0–18}	{0–18}	{0–18}	{0–11}	{0–18}
busfndrs	0.6927	0.6624	0.6839	0.6357	0.7305
	(1.015)	(1.460)	(1.018)	(0.9760)	(0.9326)
	{0–8}	{0–18}	{0–6}	{0–5}	{0–5}
acadfndrs	1.078	1.325	1.181	1.121	1.142
	(1.280)	(1.064)	(1.360)	(1.226)	(1.205)
	{0–8}	{0–7}	{0–6}	{0–5}	{0–7}
numsvyd	4.106	2.194	5.258	3.379	2.965
	(6.046)	(2.583)	(7.645)	(4.708)	(3.932)
	{1–31}	{1–27}	{1–31}	{1–27}	{1–27}
phIItsvy	13.96	4.798	19.86	10.68	9.156
	(27.46)	(10.51)	(35.14)	(21.68)	(18.77)
	{1–127}	{1–120}	{1–127}	{1–120}	{1–120}

(continued)

Table A.2 (continued)

Variable category	DoD (n = 755)[a]	NIH (n = 391)[a]	NASA (n = 155)[a]	DOE (n = 140)[a]	NSF (n = 141)[a]
percrevssbir	35.91	34.32	30.97	33.22	35.70
	(32.11)	(36.29)	(27.80)	(33.87)	(33.97)
	{0–100}	{0–100}	{0–100}	{0–100}	{0–100}
spinoffs	0.4530	0.2864	0.6065	0.5000	0.4610
	(1.143)	(0.7811)	(1.430)	(1.214)	(1.156)
	{0–6}	{0–6}	{0–6}	{0–6}	{0–6}
SBIR research variables					
awardamt	718,999.6	653,972.2	567,243.1	682,761.3	375,938.1
	(359,360.4)	(211,411.5)	(73,385.9)	(104,086.4)	(74,389.2)
	{69,673–6,190,970}	{14,834–1,644,022}	{350,000–1,125,000}	{347,681–900,000}	{213,271–500,001}
prjage	7.464	7.233	8.084	7.836	6.894
	(2.810)	(2.586)	(2.923)	(2.732)	(2.466)
	{4–13}	{4–13}	{4–13}	{4–13}	{4–13}
univpartcptn	0.2581	0.5288	0.3007	0.3824	0.5111
	(0.4379)	(0.4998)	(0.4600)	(0.4878)	(0.5017)
	{0/1)	{0/1}	{0/1}	{0/1}	{0/1}
	n = 744	n = 382	n = 153	n = 136	n = 135
sbir-r&d-to-total-r&d	40.85	40.59	38.01	37.26	39.57
	(30.98)	(36.34)	(30.20)	(34.69)	(33.67)
	{0–100}	{0–100}	{0–100}	{0–100}	{0–100}
sbirpatents	8.315	2.982	12.75	7.779	7.830
	(21.02)	(6.586)	(27.08)	(13.19)	(13.44)
	{0–125}	{0–66}	{0–125}	{0–66}	{0–66}

rltdphI	2.527	3.043	2.355	3.051	2.390
	(3.772)	(5.392)	(2.535)	(5.904)	(2.059)
	{1–83}	{1–45}	{1–20}	{1–66}	{1–13}
				n = 138	
rltdphII	1.870	2.141	1.748	1.935	1.794
	(1.629)	(3.387)	(1.527)	(1.626)	(1.365)
	{1–29}	{1–29}	{1–13}	{1–13}	{1–10}
				n = 138	
prevphII	8.551	2.120	13.75	7.107	4.355
	(26.69)	(7.241)	(36.56)	(18.49)	(12.57)
	{0–193)	{0–79}	{0–222}	{0–116}	{0–90}
Market variables					
fedacq	0.1205	0.01790	0.05806	0.05714	0.04965
	(0.3258)	(0.1328)	(0.2346)	(0.2329)	(0.2180)
	{0/1}	{0/1}	{0/1}	{0/1}	{0/1}
lateeval	0.1444	0.0486	0.09032	0.1286	0.1064
	(0.3517)	(0.2153)	(0.2876)	(0.3359)	(0.3094)
	{0/1}	{0/1}	{0/1}	{0/1}	{0/1}
Commercial-type variables					
nocom	0.02119	0.01279	0.006	0.02143	0.01418
	(0.1441)	(0.1125)	(0.080)	(0.1453)	(0.1187)
	{0/1}	{0/1}	{0/1}	{0/1}	{0/1}
software	0.2159	0.2762	0.1419	0.1357	0.2411
	(0.4117)	(0.4477)	(0.3501)	(0.3437)	(0.4293)
	{0/1}	{0/1}	{0/1}	{0/1}	{0/1}

(continued)

Table A.2 (continued)

Variable category	DoD (n = 755)[a]	NIH (n = 391)[a]	NASA (n = 155)[a]	DOE (n = 140)[a]	NSF (n = 141)[a]
hardware	0.3974	0.2558	0.4323	0.4143	0.3972
	(0.4897)	(0.4368)	(0.4970)	(0.4944)	(0.4911)
	{0/1}	{0/1}	{0/1}	{0/1}	{0/1}
process	0.1563	0.09719	0.1355	0.2357	0.2128
	(0.3634)	(0.2966)	(0.3433)	(0.4260)	(0.4107)
	{0/1}	{0/1}	{0/1}	{0/1}	{0/1}
service	0.1205	0.1125	0.1226	0.1571	0.1560
	(0.3258)	(0.3164)	(0.3290)	(0.3652)	(0.3642)
	{0/1}	{0/1}	{0/1}	{0/1}	{0/1}
drug	—	0.01790	—	—	—
		(0.1328)			
		{0/1}			
biologic	—	0.03836	—	—	—
		(0.1923)			
		{0/1}			
research	0.1046	0.2634	0.1484	0.1071	0.1560
	(0.3063)	(0.4411)	(0.3566)	(0.3104)	(0.3642)
	{0/1}	{0/1}	{0/1}	{0/1}	{0/1}
education	0.1457	0.1458	0.03226	0.02857	0.09220
	(0.1199)	(0.3533)	(0.1773)	(0.1672)	(0.2903)
	{0/1}	{0/1}	{0/1}	{0/1}	{0/1}

other	0.06225	0.07673	0.04516	0.05714	0.05674
	(0.2418)	(0.2665)	(0.2083)	(0.2329)	(0.2322)
	{0/1}	{0/1}	{0/1}	{0/1}	{0/1}
Geographic variables					
northeast	0.3099	0.3223	0.2903	0.2786	0.3404
	(0.4628)	(0.4679)	(0.4554)	(0.4499)	(0.4755)
	{0/1}	{0/1}	{0/1}	{0/1}	{0/1}
south	0.2490	0.2558	0.2323	0.1643	0.1560
	(0.4327)	(0.4368)	(0.4236)	(0.3719)	(0.3642)
	{0/1}	{0/1}	{0/1}	{0/1}	{0/1}
midwest	0.1250	0.1841	0.09677	0.1429	0.1560
	(0.3258)	(0.3881)	(0.2966)	(0.3512)	(0.3642)
	{0/1}	{0/1}	{0/1}	{0/1}	{0/1}
west	0.3205	0.2379	0.3806	0.4143	0.3475
	(0.4670)	(0.4263)	(0.4871)	(0.4944)	(0.4779)
	{0/1}	{0/1}	{0/1}	{0/1}	{0/1}
Intellectual property variables					
patents	0.9073	1.038	0.3226	0.8000	1.043
	(5.292)	(2.218)	(0.6338)	(1.619)	(1.599)
	{0–100}	{0–25}	{0–3}	{0–13}	{0–9}
copyrights	0.0715	0.6368	0.04516	0.1429	0.2979
	(0.4218)	(2.922)	(0.4163)	(0.5571)	(1.423)
	{0–6}	{0–49}	{0–5}	{0–5}	{0–11}
trademarks	0.2066	0.3990	0.07742	0.1857	0.2057
	(1.376)	(0.8130)	(0.3329)	(0.5442)	(0.5414)
	{0–35}	{0–7}	{0–3}	{0–3}	{0–3}

(continued)

Table A.2 (continued)

Variable category	DoD (n = 755)[a]	NIH (n = 391)[a]	NASA (n = 155)[a]	DOE (n = 140)[a]	NSF (n = 141)[a]
publications	1.299 (3.330) {0–50}	2.609 (9.889) {0–165}	1.413 (3.199) {0–30}	1.350 (2.056) {0–11}	1.702 (4.154) {0–41}
Funding variables					
non-sbir-federal-to-total-investment	0.07676 (0.1830) {0–0.9869}	0.02170 (0.1023) {0–0.8853} n = 389	0.03892 (0.1285) {0–0.7121}	0.05651 (0.1570) {0–0.8897} n = 137	0.07531 (0.1877) {0–0.9464} n = 137
U.S.-private-venture-capital-to-total-investment	0.01633 (0.1051) {0–0.9723}	0.01337 (0.08459) {0–0.7778} n = 389	—	—	0.01200 (0.07515) {0–0.6723} n = 137
foreign-private-investment-to-total-investment	0.003627 (0.03303) {0–0.4643}	0.006172 (0.04089) {0–0.3843} n = 389	0.001806 (0.02006) {0–0.2478}	0.008724 (0.08344) {0–0.9535} n = 137	0.009253 (0.06368) {0–0.6035} n = 137
other-private-equity-investment-to-total-investment	0.02035 (0.09850) {0–0.8889}	0.02306 (0.1007) {0–0.8532} n = 389	0.003654 (0.03014) {0–0.3571}	0.02078 (0.08809) {0–0.4828} n = 137	0.04682 (0.1526) {0–0.9091} n = 137
other-domestic-private-firm-investment-to-total-investment	0.02769 (0.09864) {0–0.7834}	0.01849 (0.09692) {0–0.8529} n = 389	0.02364 (0.08460) {0–0.4568}	0.06162 (0.1482) {0–0.7696} n = 137	0.05487 (0.1380) {0–0.7245} n = 137

state-or-local-government-funding-to-total-investment	0.004111 (0.02722) {0–0.2811}	0.007888 (0.04751) {0–0.6471} *n* = 389	0.007415 (0.06995) {0–0.8000}	0.001925 (0.01489) {0–0.1533} *n* = 137	0.01986 (0.07613) {0–0.4778} *n* = 137
colleges-or-universities-funding-to-total-investment	0.001057 (0.01694) {0–0.4000}	0.0004164 (0.003475) {0–0.04243} *n* = 389	0.001161 (0.01025) {0–0.09960}	—	—
own-firm-funding-to-total-investment	0.05091 (0.1222) {0–0.9474}	0.08527 (0.1618) {0–0.8813} *n* = 389	0.07886 (0.1752) {0–0.8096}	0.07310 (0.1525) {0–0.8571} *n* = 137	0.06419 (0.1313) {0–0.6281} *n* = 137
personal-funds-to-total-investment	0.008424 (0.04517) {0–0.6263}	0.01875 (0.06519) {0–0.5715} *n* = 389	0.003100 (0.01796) {0–0.1799}	0.004356 (0.02324) {0–0.2107} *n* = 137	0.02767 (0.08308) {0–0.5334} *n* = 137
Commercial agreement variables					
licensing agreement(s)	0.3333 (0.4718) {0/1} *n* = 594	0.3787 (0.4858) {0/1} *n* = 338	0.2411 (0.4297) {0/1} *n* = 112	0.3596 (0.4820) {0/1} *n* = 114	0.4711 (0.5012) {0/1} *n* = 121
sale of company	0.06397 (0.2449) {0/1} *n* = 594	0.05030 (0.2189) {0/1} *n* = 338	0.03571 (0.1864) {0/1} *n* = 112	0.03509 (0.1848) {0/1} *n* = 114	0.06612 (0.2495) {0/1} *n* = 121

(continued)

Table A.2 (continued)

Variable category	DoD (n = 755)[a]	NIH (n = 391)[a]	NASA (n = 155)[a]	DOE (n = 140)[a]	NSF (n = 141)[a]
partial sale of company	0.05556	0.06509	0.02679	0.05263	0.09917
	(0.2293)	(0.2470)	(0.1622)	(0.2243)	(0.3001)
	{0/1}	{0/1}	{0/1}	{0/1}	{0/1}
	n = 594	n = 338	n = 112	n = 114	n = 121
sale of technology rights	0.1566	0.1302	0.09821	0.1404	0.2645
	(0.3637)	(0.3370)	(0.2989)	(0.3489)	(0.4429)
	{0/1}	{0/1}	{0/1}	{0/1}	{0/1}
	n = 594	n = 338	n = 112	n = 114	n = 121
company merger	0.03704	0.03254	0.0	0.01754	0.06612
	(0.1890)	(0.1777)	(0.0)	(0.1319)	(0.2495)
	{0/1}	{0/1}	{0/0}	{0/1}	{0/1}
	n = 594	n = 338	n = 112	n = 114	n = 121
joint venture agreement(s)	0.1330	0.1243	0.07143	0.1140	0.1488
	(0.3399)	(0.3304)	(0.2587)	(0.3193)	(0.3573)
	{0/1}	{0/1}	{0/1}	{0/1}	{0/1}
	n = 594	n = 338	n = 112	n = 114	n = 121
marketing/ distribution agreement(s)	0.2205	0.3402	0.1696	0.2018	0.3058
	(0.4150)	(0.4745)	(0.3770)	(0.4031)	(0.4627)
	{0/1}	{0/1}	{0/1}	{0/1}	{0/1}
	n = 594	n = 338	n = 112	n = 114	n = 121
manufacturing agreement(s)	0.1498	0.1243	0.08929	0.1579	0.2066
	(0.3572)	(0.3304)	(0.2864)	(0.3663)	(0.4066)
	{0/1}	{0/1}	{0/1}	{0/1}	{0/1}
	n = 594	n = 338	n = 112	n = 114	n = 121

r&d agreement(s)	0.2828	0.2751	0.2589	0.2456	0.3719
	(0.4508)	(0.4473)	(0.4400)	(0.4324)	(0.4853)
	{0/1}	{0/1}	{0/1}	{0/1}	{0/1}
	$n = 594$	$n = 338$	$n = 112$	$n = 114$	$n = 121$
customer alliance(s)	0.2761	0.1805	0.2054	0.2018	0.3058
	(0.4474)	(0.3852)	(0.4058)	(0.4031)	(0.4627)
	{0/1}	{0/1}	{0/1}	{0/1}	{0/1}
	$n = 594$	$n = 338$	$n = 112$	$n = 114$	$n = 121$
other	0.0387	0.04438	0.05357	0.02632	0.04132
	(0.1931)	(0.2062)	(0.2262)	(0.1608)	(0.1999)
	{0/1}	{0/1}	{0/1}	{0/1}	{0/1}
	$n = 594$	$n = 338$	$n = 112$	$n = 114$	$n = 121$

NOTE: — = data not available.

[a] Smaller sample sizes (n) due to missing observations are noted in the table.

SOURCE: Authors' calculations.

Appendix B
Technical Appendix

Table B.1 Construction of the Random Sample of Projects from Tables 3.1 and 3.2: Phase II Projects (11,214 in total) for All Agencies, 1992–2001

Data reduction process	DoD	NIH	NASA	DOE	NSF
Total Phase II projects, 1992–2001	5,650	2,497	1,488	808	771
Total Phase II projects for firms with 1 project	1,048	728	254	136	204
Surveyed Phase II projects for firms with 1 project	1,047	728	254	136	204
Did not respond	809	540	212	96	139
Did respond	238	188	42	40	65
Randomly sampled	1,047	728	254	136	204
Sample weight	1.00[a]	1.00[b]	1.00[c]	1.00[d]	1.00[e]
Total Phase II projects for firms with 2 projects	651	399	159	86	83
Surveyed Phase II projects for firms with 2 projects	622	395	155	85	82
Did not respond	470	285	127	62	57
Did respond	152	110	28	23	25
Randomly sampled	622	395	155	85	82
Sample weight	1.05[f]	1.01[g]	1.03[h]	1.01[i]	1.01[j]
Total Phase II projects for firms with > 2 projects	3,951	1,370	1,075	586	484
Surveyed Phase II projects for firms with > 2 projects	1,386	555	370	218	171
Did not respond	856	357	259	124	99
Did respond	530	198	111	94	72
Randomly sampled	1,357[k]	554[l]	366[m]	215[n]	170[o]
Sample weight	2.91[p]	2.47[q]	2.94[r]	2.73[s]	2.85[t]

NOTES:
Total DoD projects = 1,048 + 651 + 3,951 = 5,650.
Total DoD surveyed = 1,047 + 622 + 1,386 = 3,055.
Total DoD randomly surveyed = 1,047 + 622 + 1,386 – 29 = 3026.
The 238 + 152 + 530 = 920 responses minus the 29 cases added to the sample = 891 responses from the random sample.
Total NIH projects = 728 + 399 + 1,370 = 2,497.
Total NIH surveyed = 728 + 395 + 555 = 1,678.
Total NIH randomly surveyed = 728 + 395 + 554 = 1,677.
The 188 + 110 + 198 = 496 responses minus the 1 case added to the sample = 495 responses from the random sample.
Total NASA projects = 254 + 159 + 1,075 = 1,488.
Total NASA surveyed = 254 + 155 + 370 = 779.
Total NASA randomly surveyed = 254 + 155 + 366 = 775.
The 42 + 28 + 111 = 181 responses minus the 4 cases added to the sample = 177 responses from the random sample.
Total DOE projects = 136 + 86 + 586 = 808.
Total DOE surveyed = 136 + 85 + 218 = 439.
Total DOE randomly surveyed = 136 + 85 + 215 = 436.
The 40 + 23 + 94 = 157 responses minus the 3 cases added to the sample = 154 responses from the random sample.
Total NSF projects = 204 + 83 + 484 = 771.
Total NSF surveyed = 204 + 82 + 171 = 457.
Total NSF randomly surveyed = 204 + 82 + 170 = 456.
The 65 + 25 + 72 = 162 responses minus the 1 case added to the sample = 161 responses from the random sample.
[a] 1,047/1,048 = 0.9990, and 1/0.9990 = 1.00.
[b] 728/728 = 1, and 1/1 = 1.
[c] 254/254 = 1, and 1/1 = 1.
[d] 136/136 = 1, and 1/1 = 1.
[e] 204/204 = 1, and 1/1 = 1.
[f] 622/651 = 0.9555, and 1/0.955 = 1.05.
[g] 395/399 = 0.9900, and 1/0.9900 = 1.01.
[h] 155/159 = 0.9748, and 1/0.9748 = 1.03.

(continued)

Table B.1 (continued)

[i] 85/86 = 0.9884, and 1/0.9884 = 1.01.

[j] 82/83 = 0.9880, and 1/0.9880 = 1.01.

[k] After taking the random sample, 5 projects were added to the sample at the request of the firms receiving the awards, and then 24 projects were added by the NRC research team to ensure that known "big successes" (over $10 million in sales and subsequent investments) were in the sample. Hence, 1,386 − 5 − 24 = 1,357.

[l] After taking the random sample, 1 project was added to the sample at the request of the firm receiving the award. Hence, 555 − 1 = 554.

[m] After taking the random sample, 4 projects were added by the NRC research team to ensure that known "big successes" (over $10 million in sales and subsequent investments) were in the sample. Hence, 370 − 4 = 366.

[n] After taking the random sample, 1 project was added to the sample at the request of the firm receiving the award, and then 2 projects were added by the NRC research team to ensure that known "big successes" (over $10 million in sales and subsequent investments) were in the sample. Hence, 218 − 3 = 215.

[o] After taking the random sample, 1 project was added by the NRC research team to ensure that known "big successes" (over $10 million in sales and subsequent investments) were in the sample. Hence, 171 − 1 = 170.

[p] 1,357/3,951 = 0.3435, and 1/0.3435 = 2.91.

[q] 554/1370 = 0.4044, and 1/0.4044 = 2.47.

[r] 366/1075 = 0.3405, and 1/0.3405 = 2.94.

[s] 215/586 = 0.3669, and 1/0.3669 = 2.73.

[t] 170/484 = 0.3512, and 1/0.3512 = 2.85.

SOURCE: Authors' calculations.

Table B.2 Negative Binomial Regression Results from Equation (4.2)[a] for the Larger Samples

	Coefficient, by agency (robust standard error)				
Variable	DoD	NIH	NASA	DOE	NSF
SBIR research variables					
univpartcptn	0.275	−0.437	—[d]	−0.227	−0.193
	(0.159)****	(0.164)******		(0.310)	(0.313)
sbir-r&d-to-total-r&d	0.00384	0.0110	−0.00426	−0.00389	0.00566
	(0.00239)***	(0.00244)*******	(0.00441)	(0.00427)	(0.00486)*
sbirpatents	−0.00599	−0.0152	0.00841	−0.0199	−0.00877
	(0.00323)****	(0.0119)**	(0.00479)****	(0.0152)**	(0.00736)*
rltdphII	0.159	−0.0435	0.0855	0.190	0.278
	(0.0723)*****	(0.0340)**	(0.0757)	(0.0735)******	(0.117)*****
Market variables					
fedacq	0.433	0.262	−0.0577	0.594	0.837
	(0.159)******	(0.529)	(0.467)	(0.345)****	(0.519)***
lateeval	0.212	−0.241	−0.367	−0.701	−0.0836
	(0.185)*	(0.552)	(0.349)	(0.596)*	(0.312)
Commercial-type variables					
nocom	0.158	−0.489	—[e]	—	1.41
	(0.433)	(0.745)			(0.545)******
software	0.778	0.194	0.274	0.554	0.143
	(0.181)*******	(0.206)	(0.277)	(0.417)**	(0.415)

(continued)

Table B.2 (continued)

Variable	Coefficient, by agency (robust standard error)				
	DoD	NIH	NASA	DOE	NSF
hardware	0.712	0.141	0.801	0.684	0.568
	(0.142)*******	(0.197)	(0.293)******	(0.376)****	(0.377)***
process	0.501	0.192	0.701	−0.0657	0.776
	(0.153)*******	(0.218)	(0.378)****	(0.369)	(0.326)*****
service	0.461	−0.0776	0.639	−0.501	0.980
	(0.284)***	(0.240)	(0.333)****	(0.327)***	(0.326)******
drug	—	1.35	—[e]	—	—
		(0.470)******			
biologic	—	−0.246	—[e]	—	—
		(0.308)			
research	−0.267	0.204	0.836	−0.494	0.0379
	(0.181)***	(0.160)*	(0.354)*****	(0.528)	(0.413)
education	−1.15	0.478	−2.10	−1.25	−0.950
	(0.434)******	(0.211)*****	(0.701)******	(0.388)*******	(0.604)***
other	0.280	0.361	0.227	1.34	−0.280
	(0.203)**	(0.334)	(0.665)	(0.524)******	(0.607)
Geographic variables					
northeast	−0.357	0.242	−1.31	−0.00212	0.769
	(0.176)*****	(0.257)	(0.363)*******	(0.337)	(0.338)*****
south	0.128	0.246	0.372	0.470	0.0388
	(0.204)	(0.254)	(0.297)*	(0.553)	(0.472)
midwest	−0.489	0.309	0.740	−0.166	0.264
	(0.249)*****	(0.237)**	(0.535)**	(0.527)	(0.373)

Intellectual property variables					
patents	0.0307 (0.00592)*******	0.0412 (0.0344)*	0.299 (0.243)*	−0.108 (0.0957)	0.230 (0.127)****
copyrights	0.0403 (0.0743)	−0.00338 (0.0132)	0.415 (0.0888)*******	0.626 (0.237)******	0.0271 (0.0981)
trademarks	−0.0264 (0.0279)	0.376 (0.122)******	0.267 (0.820)	−0.106 (0.243)	0.377 (0.361)
publications	0.0173 (0.0191)	−0.00140 (0.00343)	−0.00160 (0.0320)	0.0422 (0.0625)	0.106 (0.0338)******
Funding variables					
non-SBIR-federal-to-total-investment	1.35 (0.278)*******	1.59 (0.709)*****	1.71 (1.32)**	0.278 (0.776)	0.613 (0.865)
U.S.-private-venture-capital-to-total-investment	0.207 (0.485)	3.31 (1.01)*******	—[e]	—	−2.57 (1.12)*****
foreign-private-investment-to-total-investment	2.73 (0.967)******	2.94 (1.19)*****	−4.53 (1.93)*****	3.25 (1.28)******	−2.27 (0.940)*****

(continued)

Table B.2 (continued)

Variable	Coefficient, by agency (robust standard error)				
	DoD	NIH	NASA	DOE	NSF
other-private-equity-investment-to-total-investment	1.70 (0.579)******	1.25 (0.540)*****	−6.80 (3.90)****	−0.205 (2.41)	−0.831 (0.779)
other-domestic-private-firm-investment-to-total-investment	−0.380 (0.544)	1.19 (0.736)***	−1.25 (1.00)*	2.34 (0.755)******	0.224 (0.906)
state-or-local-government-funding-to-total-investment	−0.700 (1.69)	2.85 (0.706)*******	−3.58 (1.63)*****	9.05 (8.90)	−0.733 (1.69)
colleges-or-universities-funding-to-total-investment	−0.714 (1.05)	−4.46 (9.90)	−8.92 (6.51)**	—	—

own-firm-funding-to-total-investment	1.16 (0.365)******	3.24 (0.476)*******	0.253 (0.668)	1.55 (0.534)******	1.45 (0.926)***
personal-funds-to-total-investment	1.54 (0.974)***	1.76 (0.943)****	10.3 (7.06)***	4.14 (9.86)	3.19 (2.31)**
Constant	−3.03 (0.243)*******	−2.89 (0.276)*******	−2.92 (0.412)*******	−2.81 (0.505)*******	−3.71 (0.411)*******
Auxiliary parameter: alpha[b]	1.10 (0.170)	0.761 (0.138)	0.878 (0.245)	5.91×10^{-7} (7.86×10^{-6})	1.05 (0.313)
n	744	382	155	134	135
Log pseudo likelihood	−2526.97	−971.83	−431.01	−360.36	−397.73
Wald chi-squared (df)[c]	1,052.74*******	344.86*******	137.99*******	1,868.76*******	247.27*******

NOTE: — = data not available. Exposure = *prjage*. Significance levels (two-tailed excepting chi-squared): ******* = 0.001; ****** = 0.01; ***** = 0.05; **** = 0.10; *** = 0.15; ** = 0.20; * = 0.25. (df) = degrees of freedom.

DoD

[a] Estimation with probability weights (also called sampling weights) and standard errors adjusted for 478 clusters by firm.

[b] Significantly different from 0 (the 95% confidence interval for alpha is 0.811 to 1.49), so the negative binomial rather than the Poisson estimator is appropriate.

[c] df = 30.

NIH

[a] Estimation with probability weights (also called sampling weights) and standard errors adjusted for 282 clusters by firm.

(continued)

Table B.2 (continued)

[b] Significantly different from 0 (the 95% confidence interval for alpha is 0.533 to 1.09), so the negative binomial rather than the Poisson estimator is appropriate.

[c] df = 32.

NASA

[a] Estimation with probability weights (also called sampling weights) and standard errors adjusted for 123 clusters by firm.

[b] Significantly different from 0 (the 95% confidence interval for alpha is 0.507 to 1.52), so the negative binomial rather than the Poisson estimator is appropriate.

[c] df = 27.

[d] *univpartcptn* not included because including the variable—the only one that constrained the size of the sample—precluded using the full sample of respondents, and the variable did not have a significant effect in the subsample for which the model including it could be estimated.

[e] Occasionally variables are excluded because there are insufficient nonzero observations in the context of the many variables with small numbers of nonzero observations for estimation of the model statistics. For example, *nocom* was excluded and the effect left in the intercept; there was just a single observation in the sample, with *nocom* equal to 1. For another example, *drug* is omitted because it equaled zero in all 155 observations for the NASA sample. These types of cases appeared in the three smaller samples—NASA, DOE, and NSF—and even in the large DoD sample for the variables *drug* and *biologic*.

DOE

[a] Estimation with probability weights (also called sampling weights) and standard errors adjusted for 104 clusters by firm.

[b] With the relatively large standard error, the 95% confidence interval for alpha ranges from 2.85×10^{-18} to 122,785.7, so the more general negative binomial rather than the Poisson estimator is appropriate. Alpha is not well estimated, but we certainly cannot assume that it is zero.

[c] df = 27.

NSF

[a] Estimation with probability weights (also called sampling weights) and standard errors adjusted for 121 clusters by firm.

[b] Significantly different from 0 (the 95% confidence interval for alpha is 0.584 to 1.88), so the negative binomial rather than the Poisson estimator is appropriate.

[c] df = 29.

SOURCE: Authors' calculations.

Table B.3 Negative Binomial Regression Results from Equation (4.2)[a] for Smaller Samples with Commercial Agreement Variables

Variable	Coefficient, by agency (robust standard error)				
	DoD	NIH	NASA	DOE	NSF
SBIR research variables					
univpartcptn	0.385	−0.476	—	−0.289	−0.341
	(0.167)*****	(0.165)******		(0.353)	(0.354)
sbir-r&d-to-total-r&d	0.00260	0.0106	−0.00484	0.000279	0.000568
	(0.00228)	(0.00250)*******	(0.00481)	(0.00477)	(0.00551)
sbirpatents	−0.00437	−0.0189	0.00922	−0.0292	−0.0265
	(0.00342)*	(0.0124)***	(0.00503)****	(0.0119)*****	(0.0109)*****
rltdphII	0.132	−0.0644	−0.0419	0.114	0.269
	(0.0730)****	(0.0316)*****	(0.131)	(0.0732)***	(0.105)******
Market variables					
fedacq	0.429	−0.124	0.109	0.647	1.50
	(0.139)******	(0.484)	(0.482)	(0.367)****	(0.723)*****
lateeval	0.178	0.0782	−0.371	− 0.480	−0.745
	(0.187)	(0.612)	(0.475)	(0.676)	(0.318)*****
Commercial-type variables					
nocom	−0.669	−0.632	—	—	0.880
	(0.433)***	(0.931)			(0.865)

(continued)

Table B.3 (continued)

Variable	Coefficient, by agency (robust standard error)				
	DoD	NIH	NASA	DOE	NSF
software	0.629	0.0433	0.131	0.127	0.283
	(0.185)*******	(0.204)	(0.314)	(0.405)	(0.393)
hardware	0.379	−0.148	0.515	0.232	−0.221
	(0.156)*****	(0.196)	(0.327)***	(0.457)	(0.366)
process	0.337	−0.111	0.706	−0.0715	1.08
	(0.152)*****	(0.211)	(0.400)****	(0.421)	(0.430)*****
service	0.275	−0.174	0.565	−0.458	0.959
	(0.264)	(0.240)	(0.352)***	(0.299)***	(0.328)******
drug	—	0.942	—	—	—
		(0.442)*****			
biologic	—	−0.572	—	—	—
		(0.318)****			
research	−0.356	0.118	0.822	−1.13	−0.0754
	(0.171)*****	(0.168)	(0.343)*****	(0.886)*	(0.403)
education	−0.15	0.342	−0.136	−1.11	−1.38
	(0.365)******	(0.219)***	(0.803)	(0.752)***	(0.638)*****
other	0.0876	0.0477	0.219	0.693	0.109
	(0.192)	(0.314)	(0.638)	(0.679)	(0.509)
Geographic variables					
northeast	−0.142	0.418	−2.00	0.163	1.36
	(0.175)	(0.254)****	(0.498)*******	(0.296)	(0.441)******

south	0.123 (0.195)	0.238 (0.266)	0.00812 (0.348)	0.198 (0.649)	0.0249 (0.441)
midwest	−0.344 (0.228)***	0.379 (0.232)***	−0.211 (0.370)	−0.317 (0.593)	1.05 (0.413)*****
Intellectual property variables					
patents	0.0285 (0.00663)*******	0.0508 (0.0357)**	0.397 (0.249)***	−0.0739 (0.0978)	0.217 (0.104)*****
copyrights	0.0623 (0.0704)	−0.000269 (0.0146)	0.688 (0.219)******	0.609 (0.228)******	−0.178 (0.0969)****
trademarks	−0.00543 (0.0303)	0.381 (0.127)******	0.0434 (0.488)	−0.373 (0.335)	0.292 (0.455)
publications	0.00657 (0.0187)	0.000237 (0.00322)	0.00319 (0.0358)	0.0591 (0.0624)	0.0622 (0.0312)*****
Funding variables					
non-SBIR-federal-to-total-investment	1.18 (0.260)*******	1.48 (0.694)*****	1.48 (1.59)	−0.155 (0.720)	−1.62 (0.887)****
U.S.-private-venture-capital-to-total-investment	0.113 (0.501)	3.23 (1.03)******	—	—	−2.31 (1.89)*
foreign-private-investment-to-total-investment	1.82 (0.935)****	3.89 (1.46)******	−3.67 (4.30)	3.45 (1.24)******	−2.65 (1.08)*****
other-private-equity-investment-to-total-investment	1.26 (0.518)*****	1.17 (0.519)*****	− 4.51 (6.03)	0.629 (2.79)	−2.44 (0.955)*****

(continued)

Table B.3 (continued)

Variable	Coefficient, by agency (robust standard error)				
	DoD	NIH	NASA	DOE	NSF
other-domestic-private-firm-investment-to-total-investment	−0.204 (0.513)	1.19 (0.925)**	0.674 (1.43)	2.30 (0.806)******	−1.21 (0.809)***
state-or-local-government-funding-to-total-investment	−1.16 (1.61)	2.67 (0.699)*******	—	0.125 (9.07)	−2.85 (1.45)*****
colleges-or-universities-funding-to-total-investment	−0.876 (0.976)	1.95 (8.69)	−11.9 (9.64)*	—	—
own-firm-funding-to-total-investment	1.02 (0.372)******	3.12 (0.478)*******	1.11 (1.05)	1.82 (0.670)******	0.0191 (1.14)
personal-funds-to-total-investment	0.142 (0.910)	2.06 (0.825)*****	4.74 (8.63)	11.7 (6.95)****	0.231 (1.90)
Commercial agreement variables					
licensing agreement(s)	−0.309 (0.136)*****	−0.120 (0.226)	−0.931 (0.540)****	0.335 (0.428)	1.21 (0.467)******
sale of company	0.394 (0.271)***	−1.06 (0.461)*****	0.117 (0.669)	0.165 (0.890)	−0.00668 (0.736)

partial sale of company	−0.228 (0.222)	0.207 (0.360)	−0.0496 (0.834)	−0.121 (0.732)	0.368 (0.627)
sale of technology rights	0.198 (0.197)	−0.410 (0.272)***	−1.27 (0.803)***	−0.942 (0.858)	−0.515 (0.364)**
company merger	0.492 (0.321)***	0.683 (0.598)	—	—	0.184 (0.950)
joint venture agreement(s)	−0.185 (0.182)	0.530 (0.271)****	1.94 (0.783)*****	0.558 (0.519)	−1.25 (0.579)*****
marketing/distribution agreement(s)	0.305 (0.123)*****	0.146 (0.193)	0.110 (0.456)	−0.112 (0.370)	0.967 (0.398)*****
manufacturing agreement(s)	0.249 (0.161)***	−0.0715 (0.282)	1.33 (0.950)**	−0.379 (0.428)	0.699 (0.412)****
r&d agreement(s)	−0.0294 (0.140)	0.0867 (0.178)	−1.00 (0.422)*****	0.852 (0.369)*****	−0.0163 (0.330)
customer alliance(s)	−0.160 (0.136)*	0.176 (0.184)	−1.26 (0.506)*****	0.188 (0.339)	0.565 (0.344)****
other	−0.464 (0.266)****	−0.759 (0.347)*****	−1.51 (0.680)*****	0.228 (0.505)	−2.69 (0.691)*******
Constant	−2.56 (0.278)*******	−2.56 (0.303)*******	−1.81 (0.519)*******	−2.41 (0.559)*******	−3.33 (0.480)*******
Auxiliary parameter: alpha[b]	0.862 (0.151)	0.647 (0.126)	0.375 (0.184)	1.23×10^{-8} (7.26×10^{-9})	0.533 (0.151)
n	552	307	109	104	106
Log pseudo likelihood	−2,144.89	−829.37	−318.70	−295.24	−328.84
Wald chi-squared (df)[c]	1,522.83*******	366.91*******	169.14*******	3,456.31*******	367.60*******

(continued)

Table B.3 (continued)

NOTE: — = data not available. Exposure = *prjage*. Significance levels (two-tailed excepting chi-squared): ******* = 0.001; ****** = 0.01; ***** = 0.05; **** = 0.10; *** = 0.15; ** = 0.20; * = 0.25. (df) = degrees of freedom.

DoD

[a] Estimation with probability weights (also called sampling weights) and standard errors adjusted for 392 clusters by firm.

[b] Significantly different from 0 (the 95% confidence interval for alpha is 0.612 to 1.21), so the negative binomial rather than the Poisson estimator is appropriate.

[c] df = 41.

NIH

[a] Estimation with probability weights (also called sampling weights) and standard errors adjusted for 240 clusters by firm.

[b] Significantly different from 0 (the 95% confidence interval for alpha is 0.443 to 0.947), so the negative binomial rather than the Poisson estimator is appropriate.

[c] df = 43.

NASA

[a] Estimation with probability weights (also called sampling weights) and standard errors adjusted for 94 clusters by firm.

[b] Significantly different from 0 (the 95% confidence interval for alpha is 0.144 to 0.980), so the negative binomial rather than the Poisson estimator is appropriate.

[c] df = 36.

DOE

[a] Estimation with probability weights (also called sampling weights) and standard errors adjusted for 86 clusters by firm.

[b] Because alpha is essentially 0 (the 95% confidence interval for alpha is 3.88×10^{-9} to 3.91×10^{-8}), the Poisson estimator would do just as well as the negative binomial. The Poisson distribution is a special case of the negative binomial distribution with alpha equal to zero.

[c] df = 37.

NSF

[a] Estimation with probability weights (also called sampling weights) and standard errors adjusted for 95 clusters by firm.

[b] Significantly different from 0 (the 95% confidence interval for alpha is 0.306 to 0.929), so the negative binomial rather than the Poisson estimator is appropriate.

[c] df = 40.

SOURCE: Authors' calculations.

Table B.4 Percentage Change in *retainees*, for Statistically Significant Changes, Given Change in the Explanatory Variable, Other Things Being the Same

Explanatory variable	DoD	NIH	NASA	DOE	NSF
SBIR research variables					
univpartcptn	47.0	−37.9			
sbir-r&d-to-total-r&d		11.2			
sbirpatents				−2.9	−2.6
rltdphII		−6.2			30.9
Market variables					
fedacq	53.6				348.0
lateeval					−52.5
Commercial-type variables					
software	87.6				
hardware	46.1				
process	40.1				194.0
service					161.0
drug		157.0			
research	−30.0		128.0		
education	−68.3				−74.8
Geographic variables					
northeast			−86.5		290.0
midwest					186.0
Intellectual property variables					
patents	2.9				24.2
copyrights			99.0	83.9	
trademarks		46.4			
publications					6.4
Funding variables					
non-SBIR-federal-to-total-investment	12.5	16.0			
U.S.-private-venture-capital-to-total-investment		38.1			
foreign-private-investment-to-total-investment		47.6		41.2	−23.3
other-private-equity-investment-to-total-investment	13.4	12.4			−21.7

(continued)

Table B.4 (continued)

Explanatory variable	DoD	NIH	NASA	DOE	NSF
other-domestic-private-firm-investment-to-total-investment				25.9	
state-or-local-government-funding-to-total-investment		30.6			−24.8
own-firm-funding-to-total-investment	10.7	36.6		20.0	
personal-funds-to-total-investment		22.9			
Commercial agreement variables					
licensing agreement(s)	−26.6				235.0
sale of company		−65.4			
joint venture agreement(s)			596.0		−71.3
marketing/distribution agreement(s)	35.7				163.0
r&d agreement(s)			−63.2	134.0	
customer alliance(s)			−71.6		
other		−53.2	−77.9		−93.2

NOTE: Blank = not applicable. The qualifying phrase 'Given Change in the Explanatory Variable' in the table's title means a change of 1.0 unit in the explanatory variable. There are two exceptions. The first is *sbir-r&d-to-total-r&d,* which is a percentage and ranges from 0 to 100. For this table we consider the impact of changing it by 10 units—i.e., an increase of 10 percentage points. A second exception is made for the funding variables. These are measured as proportions and are observed in the range between 0 and 1. For this table we consider the impact of changing them by 0.1—i.e., each funding variable is increased to take 10 percent more of the total funding. Only changes significant at the 0.05 level or higher for a two-tailed test are shown. The changes are based on the estimations in Table B.3. Holding constant all variables except x_i, given a change in x_i, the expected number of *retainees* relative to the expected number without that change is the base for the natural logarithms raised to the power of the product of the coefficient for x_i and the change in x_i. For a unit change in x_i, the expression is called the incident rate ratio and is simply the base for the natural logarithms raised to the power equal to the coefficient for variable x_i. Subtracting 1.0 from the incident rate ratio and multiplying by 100 gives the percentage change shown in this table, excepting the cases noted above, for which the change in x_i differed from 1.0 (being 10 in one case and 0.1 in the others). Two caveats: 1) To understand fully the magnitude of the effect from changing an explanatory variable, the reader should consult the Glossary of Variables and Descriptive Statistics, Table A.1, to find the units in which each of the explanatory variables is measured as well as the standard deviation for each. The percentage change shown in this table is for a one-unit change

Table B.4 (continued)

in the explanatory variable, excepting the cases noted above. Following the procedure outlined in this note, one can readily compute the percentage change for various amounts of change (for example, a standard deviation) in the explanatory variable. 2) Because only those changes at the 0.05 level of significance for a two-tailed test are shown, the additional detail in Table B.3 reveals other important effects that are not quite statistically significant at the 0.05 level but that should not be ignored for a full understanding of the factors affecting the employment impact of the SBIR program.
SOURCE: Authors' calculations.

Table B.5 Descriptive Statistics for the Complete Sample of Respondents and Nonrespondents for the Employment Growth Model and Probability of Response Model

Panel A: Complete DoD Sample: Respondents and Nonrespondents

Variable	*n*	Mean	Std. dev.	Range
awardamt	3,026	697,105.9	327,073	50,000–6,190,970
ln(*awardamt*)	3,026	13.38	0.3905	10.82–15.64
phIItsvy	3,026	7.602	16.84	1–127
ln(*phIItsvy*)	3,026	1.101	1.167	0–4.884
numsvyd	3,026	2.746	3.779	1–31
prjage	3,026	8.167	2.873	4–13

NOTE: *n* = number of respondents providing information on all variables. Of the 891 projects for which completed or partially completed responses were available (see Table B.1), 136 projects were missing one or more of the explanatory variables in the prediction model, leaving 755 (891 − 136) uncensored observations to estimate the econometric model in Table B.6. Of the 136 projects not among the 755 uncensored observations but among the 891 with complete or partially complete responses, there were 104 that did not provide a response about their employment—the focal variable of this study—and therefore for purposes of the econometric model are in the sample as censored observations. Among the 136 projects, there were 32 cases where the respondents *did* provide the dependent variable (*empt* and hence its natural logarithm) for the employment model but for which one or more of the explanatory variables were missing. Those 32 cases were not censored in the sense of the 2,239 projects for which there was no response about the dependent variable here, and the econometric model is estimated with 2,994 observations (the 3,026 observations in the random sample of DoD projects minus the 32 cases that were not censored, because they responded with the dependent variable, but cannot be used in the econometric model because of missing explanatory variables). Thus, to estimate the econometric model in Table B.6, there are 2,994 observations, of which 2,239 are censored and 755 are uncensored.

Table B.5 (continued)

Panel B: Complete NIH Sample: Respondents and Nonrespondents				
Variable	n	Mean	Std. dev.	Range
awardamt	1,677	623,792.2	232,029.2	7,050–2,948,097
ln(*awardamt*)	1,677	13.26	0.4461	8.861–14.90
phIItsvy	1,677	3.537	7.212	1–120
ln(*phIItsvy*)	1,677	0.7221	0.8442	0–4.787
numsvyd	1,677	1.884	1.746	1–27
prjage	1,677	7.688	2.849	4–13

NOTE: n = number of respondents providing information on all variables. Of the 495 projects for which completed or partially completed responses were available (see Table B.1), 104 projects were missing one or more of the explanatory variables in the employment model, leaving 391 (495 − 104) uncensored observations to estimate the econometric model in Table B.6. Of the 104 projects not among the 391 uncensored observations but among the 495 with complete or partially complete responses, there were 71 that did not provide a response about their employment—the focal variable of this study—and therefore for purposes of the econometric model are in the sample as censored observations. Among the 104 projects, there were 33 cases where the respondents *did* provide the dependent variable (*empt* and hence its natural logarithm) for the employment model but for which one or more of the explanatory variables were missing. Those 33 cases were not censored in the sense of the 1,253 projects for which there was no response about the dependent variable here, and the econometric model is estimated with 1,644 observations (the 1,677 observations in the random sample of NIH projects minus the 33 cases that were not censored, because they responded with the dependent variable, but cannot be used in the econometric model because of missing explanatory variables). Thus, to estimate the econometric model in Table B.6, there are 1,644 observations, of which 1,253 are censored and 391 are uncensored.

(continued)

Table B.5 (continued)

Panel C: Complete NASA Sample: Respondents and Nonrespondents				
Variable	*n*	Mean	Std. dev.	Range
awardamt	775	554,673.6	101,767.2	1–1,982,676
ln(*awardamt*)	775	13.19	0.5176	0–14.50
phIItsvy	775	8.610	19.04	1–127
ln(*phIItsvy*)	775	1.154	1.21	0–4.844
numsvyd	775	2.919	4.169	1–31
prjage	775	9.019	2.759	4–13

NOTE: *n* = number of respondents providing information on all variables. Of the 177 projects for which completed or partially completed responses were available (see Table B.1), 22 projects were missing one or more of the explanatory variables in the employment model, leaving 155 (177 − 22) uncensored observations to estimate the econometric model in Table B.6. Of the 22 projects not among the 155 uncensored observations but among the 177 with complete or partially complete responses, there were 18 that did not provide a response about their employment—the focal variable of this study—and therefore for purposes of the econometric model are in the sample as censored observations. Among the 22 projects, there were 4 cases where the respondents *did* provide the dependent variable (*empt* and hence its natural logarithm) for the employment model but for which one or more of the explanatory variables were missing. Those 4 cases were not censored in the sense of the 616 projects for which there was no response about the dependent variable here, and the econometric model is estimated with 771 observations (the 775 observations in the random sample of NASA projects minus the 4 cases that were not censored, because they responded with the dependent variable, but cannot be used in the econometric model because of missing explanatory variables). Thus, to estimate the econometric model in Table B.6, there are 771 observations, of which 616 are censored and 155 are uncensored.

Table B.5 (continued)

Panel D: Complete DOE Sample: Respondents and Nonrespondents				
Variable	*n*	Mean	Std. dev.	Range
awardamt	436	666,741.3	108,160.5	247,838–900,000
ln(*awardamt*)	436	13.39	0.1814	12.42–13.71
phIItsvy	436	8.287	15.87	1–120
ln(*phIItsvy*)	436	1.222	1.206	0–4.787
numsvyd	436	2.933	3.562	1–27
prjage	436	8.046	2.824	4–13

NOTE: *n* = number of respondents providing information on all variables. Of the 154 projects for which completed or partially completed responses were available (see Table B.1), 14 projects were missing one or more of the explanatory variables in the employment model, leaving 140 (154 − 14) uncensored observations to estimate the model in Table B.6. Of the 14 projects not among the 140 uncensored observations but among the 154 with complete or partially complete responses, there were 8 that did not provide a response about their employment—the focal variable of this study—and therefore for purposes of the econometric model are in the sample as censored observations. Among the 14 projects, there were 6 cases where the respondents *did* provide the dependent variable (*empt* and hence its natural logarithm) for the employment model but for which one or more of the explanatory variables were missing. Those 6 cases were not censored in the sense of the 290 projects for which there was no response about the dependent variable here, and the econometric model is estimated with 430 observations (the 436 observations in the random sample of DOE projects minus the 6 cases that were not censored, because they responded with the dependent variable, but cannot be used in the econometric model because of missing explanatory variables). Thus, to estimate the econometric model in Table B.6, there are 430 observations, of which 290 are censored and 140 are uncensored.

(continued)

Table B.5 (continued)

Panel E: Complete NSF Sample: Respondents and Nonrespondents				
Variable	n	Mean	Std. dev.	Range
awardamt	456	350,413	78,787.13	6,000–500,001
ln(*awardamt*)	456	12.73	0.3209	8.700–13.12
phIItsvy	456	5.963	12.63	1–120
ln(*phIItsvy*)	456	0.9074	1.117	0–4.787
numsvyd	456	2.342	2.735	1–27
prjage	456	7.667	2.624	4–13

NOTE: n = number of respondents providing information on all variables. Of the 161 projects for which completed or partially completed responses were available (see Table B.1), 20 projects were missing one or more of the explanatory variables in the employment model, leaving 141 (161 − 20) uncensored observations to estimate the econometric model in Table B.6. Of the 20 projects not among the 141 uncensored observations but among the 161 with complete or partially complete responses, there were 10 that did not provide a response about their employment—the focal variable of this study—and therefore for purposes of the econometric model are in the sample as censored observations. Among the 20 projects, there were 10 cases where the respondents *did* provide the dependent variable (*empt* and hence its natural logarithm) for the employment model but for which one or more of the explanatory variables were missing. Those 10 cases were not censored in the sense of the 305 projects for which there was no response about the dependent variable here, and the econometric model is estimated with 446 observations (the 456 observations in the random sample of NSF projects minus the 10 cases that were not censored, because they responded with the dependent variable, but cannot be used in the econometric model because of missing explanatory variables). Thus, to estimate the econometric model in Table B.6, there are 446 observations, of which 305 are censored and 141 are uncensored.

SOURCE: Authors' calculations.

Table B.6 Results from the Firm Employment Growth Model, Equation (5.4), with Control for Response[a]

Variable	Coefficient, by agency (robust standard error)[b]				
	DoD	NIH	NASA	DOE	NSF
Regression model for ln(*empt*)					
t	0.02896	0.18547	0.15927	0.14967	0.10812
	(.0404)	(0.0558)*****	(0.0740)***	(0.0509)****	(0.0385)****
firmage × t	0.00351	−0.00853	−0.00447	−0.00408	−0.00138
	(0.0026)	(0.0053)*	(0.0042)	(0.0024)**	(0.0018)
firmage × t^2	−0.00013	0.00018	0.00004	0.00002	4.20×10^{-6}
	(0.00006)***	(0.0001)	(0.00006)	(0.00002)	(0.00002)
firmage × t^3	8.99×10^{-7}	-1.14×10^{-6}	–	–	–
	(3.46×10^{-7})****	(9.99×10^{-7})			
sbirfnd × t	0.00405	−0.0672	−0.06520	0.01489	−0.05108
	(0.0174)	(0.0166)*****	(0.0338)**	(0.0240)	(0.0263)**
otherstarts × t	−0.00152	0.00085	−0.00023	−0.00016	0.01237
	(0.0015)	(0.0048)	(0.0016)	(0.0040)	(0.0042)****
prevphII × t	0.00013	0.00091	−0.00011	0.00063	−0.00056
	(0.00008)*	(0.0006)*	(0.0002)	(0.0004)*	(0.0005)
busfndrs × t	0.00937	−0.00360	0.00180	−0.00287	0.02846
	(0.0024)*****	(0.0055)	(0.0049)	(0.0118)	(0.0193)*
acadfndrs × t	0.00530	0.01595	0.01058	0.01346	0.03217
	(0.0026)***	(0.0057)****	(0.0055)**	(0.0090)*	(0.0086)*****
lateeval × t	0.01871	−0.01982	−0.04907	−0.00136	0.01771
	(0.0122)*	(0.0384)	(0.0213)***	(0.0227)	(0.0194)

(continued)

Table B.6 (continued)

	Coefficient, by agency (robust standard error)[b]				
Variable	DoD	NIH	NASA	DOE	NSF
northeast × *t*	0.00889	−0.01398	0.01176	−0.01109	−0.02783
	(0.0010)	(0.0246)	(0.0163)	(0.0180)	(0.0153)**
south × *t*	0.03410	−0.03447	0.08332	−0.02988	−0.05539
	(0.0103)*****	(0.0245)	(0.0295)****	(0.0399)	(0.0332)**
midwest × *t*	0.02223	−0.02065	0.03441	0.05006	−0.00853
	(0.0137)*	(0.0268)	(0.0398)	(0.0326)*	(0.0293)
Constant	3.228	2.5344	3.09273	2.1893	3.1811
	(0.2143)*****	(0.3680)*****	(0.4408)*****	(0.3761)*****	(0.3433)*****
Probit model for "Response to the Survey"					
ln(*awardamt*)	−0.01558	0.13469	0.20504	0.66435	0.83227
	(0.0670)	(0.0839)*	(0.2222)	(0.4057)*	(0.5015)**
ln(*phIItsvy*)	0.13862	0.24769	0.04407	0.09363	0.18444
	(0.0482) ****	(0.0886)****	(0.0638)	(0.1122)	(0.0497)*****
numsvyd	0.03768	−0.00254	0.06455	0.01059	0.01142
	(0.0142)****	(0.0369)	(0.0181)*****	(0.0391)	(0.0139)
prjage	−0.04888	−0.03562	−0.03182	0.00253	0.00367
	(0.0100)*****	(0.0132)****	(0.0194)**	(0.0310)	(0.0359)
Constant	−0.33888	−2.4513	−3.52892	−9.4935	−11.38831
	(0.9180)	(1.0990)***	(2.97564)	(5.5423)**	(6.6563)**
Rho	−0.85080	−0.72823	−0.88485	−0.68498	−0.97940
	(0.0343)*****	(0.1005)*****	(0.0518)*****	(0.1766)***	(0.0218)*****

Sigma	1.5095 (0.1059)*****	1.2730 (0.1629)**	1.6128 (0.2330)*****	1.18964 (0.1832)	1.90361 (0.2436)*****
n	2,994	1,644	771	430	446
Censored	2,239	1,253	616	290	305
Uncensored	755	391	155	140	141
Wald chi-squared (df)	313.23 (13)*****	149.61 (13)*****	123.24 (12)*****	280.55 (12)*****	341.60 (12)*****
Log pseudo likelihood	−5,481.05	−2,237.56	−1,213.99	−904.88	−811.24
Wald chi-squared (1) test of independent equations (rho = 0)	102.51*****	18.69*****	34.33*****	6.35***	18.16*****

NOTE: — = data not available. The NASA, DOE, and NSF samples have too few uncensored observations on different ages to sensibly fit the *firmage* × t^3 term. (df) = degrees of freedom.

[a] Explanation of the total number of observations and the number of uncensored observations is given in Table B.1 and in Table B.5. Estimation with probability weights (also called sampling weights) and standard errors is adjusted for clusters by firm, respectively. The significance levels for rho and sigma are the significance levels for the estimated ancillary parameters from which rho and sigma were derived.

[b] Significance levels (two-tailed excepting chi-squares): ***** = 0.001, **** = 0.01, *** = 0.05, ** = 0.10, * = 0.15.

SOURCE: Authors' calculations.

Table B.7 Coefficients from the Performance Model in Equation (6.1)

Variable	Coefficient, by agency (robust standard error)[a]				
	DoD[b]	NIH[c]	NASA[d]	DOE[e]	NSF[f]
Firm characteristics					
spinoffs	0.146 (0.0273)*******	0.103 (0.0820)*			
percrevssbir	−0.00987 (0.00177)*******				0.00832 (0.00495)****
fedacq			0.614 (0.219)******		
SBIR research variables					
rltdphII	0.0456 (0.0208)*****	0.0235 (0.0191)*			0.175 (0.0731)*****
Intellectual property variables					
patents	0.00460 (0.00257)****	0.0656 (0.0338)****		0.107 (0.0364)******	
publications	0.0255 (0.00820)******	−0.00730 (0.00344)*****			0.0417 (0.0192)*****
copyrights			0.356 (0.101)*******		
Funding variables					
outfintoinv	0.606 (0.180)*******				
non-sbir-federal-to-total-investment					1.63 (0.661)*****

U.S.-private-venture-capital-to-total-investment		2.23 (0.651)*******			−3.91 (1.38)******
colleges-or-universities-funding-to-total-investment	−2.10 (0.546)*******	−46.4 (13.7)*******	−9.90 (6.36)**		
other-private-equity-investment-to-total-investment					1.98 (0.465)*******
other-domestic-private-firm-investment-to-total-investment					0.991 (0.651)***
state-or-local-government-funding-to-total-investment					−3.29 (1.14)******
own-firm-funding-to-total-investment		1.34 (0.470)******	0.732 (0.360)*****		2.61 (1.02)*****

(continued)

Table B.7 (continued)

	Coefficient, by agency (robust standard error)[b]				
Variable	DoD[b]	NIH[c]	NASA[d]	DOE[e]	NSF[f]
personal-funds-to-total-investment	−4.08 (1.19)*******	−1.71 (0.804)*****		−4.31 (2.17)*****	
Commercial agreement variables					
dagreement	−0.258 (0.102)*****			−0.734 (0.227)******	
sale of company		−0.748 (0.317)*****			−1.35 (0.410)*******
sale of technology rights		−0.552 (0.218)*****			
joint venture agreement(s)			−0.846 (0.296)******		
marketing/ distribution agreement(s)			0.350 (0.250)**		
r&d agreement(s)			−0.420 (0.249)****		
customer alliance(s)			0.741 (0.257)******		
other					0.957 (0.430)*****

Commercial-type variables					
nocom	0.175	−0.192		0.647	−0.370
	(0.265)	(0.438)		(0.382)****	(0.412)
software	0.103	−0.185	−0.371	0.234	0.473
	(0.120)	(0.221)	(0.233)***	(0.332)	(0.370)*
hardware	−0.0460	−0.292	−0.0866	0.342	0.941
	(0.0927)	(0.187)***	(0.216)	(0.252)**	(0.264)*******
process	−0.131	0.365	0.0661	−0.492	0.448
	(0.121)	(0.318)	(0.221)	(0.277)****	(0.295)***
service	0.0393	−0.0437	−0.249	0.472	0.280
	(0.129)	(0.216)	(0.228)	(0.278)****	(0.291)
drug		−0.0264			
		(0.336)			
biologic		0.430			
		(0.527)			
research	−0.140	0.300	0.105	−0.382	−0.901
	(0.139)	(0.232)**	(0.256)	(0.220)****	(0.292)******
education	−0.973	0.313	−1.16	−0.839	−0.374
	(0.346)******	(0.307)	(0.391)******	(0.347)*****	(0.462)
other	−0.0383	0.0810	−0.0930	0.158	−0.812
	(0.180)	(0.224)	(0.311)	(0.280)	(0.365)*****
Constant	0.495	0.00360	0.170	0.0725	−1.86
	(0.145)*******	(0.238)	(0.208)	(0.229)	(0.432)*******
n	562	312	109	106	107
F (df)[g]	15.54(17)*******	4.07(20)*******	7.15(15)*******	7.56(11)*******	6.77(19)*******
R^2	0.2164	0.1467	0.3711	0.2662	0.476

(continued)

Table B.7 (continued)

NOTE: Blank = not applicable.

[a] Significance levels (two-tailed excepting F): ******* = 0.001, ****** = 0.01, ***** = 0.05, **** = 0.10, *** = 0.15, ** = 0.20, * = 0.25.

[b] Estimation with probability weights (also called sampling weights) and standard errors adjusted for 400 clusters by firm.

[c] Estimation with probability weights (also called sampling weights) and standard errors adjusted for 245 clusters by firm.

[d] Estimation with probability weights (also called sampling weights) and standard errors adjusted for 94 clusters by firm.

[e] Estimation with probability weights (also called sampling weights) and standard errors adjusted for 88 clusters by firm.

[f] Estimation with probability weights (also called sampling weights) and standard errors adjusted for 96 clusters by firm.

[g] In this row, the first number is the F statistic, and the number in parentheses is the number of degrees of freedom for the numerator. The number of degrees of freedom for the denominator equals the number of clusters minus 1.

SOURCE: Authors' calculations.

References

Abramovitz, Moses. 1956. "Resource and Output Trends in the United States since 1870." *American Economic Review* 46(2): 5–23.

Acs, Zoltan J., and David B. Audretsch. 1990. *Innovation and Small Firms*. Cambridge, MA: MIT Press.

Arrow, Kenneth J. 1962a. "The Economic Implications of Learning by Doing." *Review of Economic Studies* 29(3): 155–173.

———. 1962b. "Economic Welfare and the Allocation of Resources for Invention." In *The Rate and Direction of Inventive Activity: Economic and Social Factors*, Richard R. Nelson, ed. Princeton, NJ: Princeton University Press, pp. 609–626.

———. 1963. *Social Choice and Individual Values*. 2d ed. New York: John Wiley and Sons.

Åstebro, Thomas. 2003. "The Return to Independent Invention: Evidence of Unrealistic Optimism, Risk Seeking or Skewness Loving?" *Economic Journal* 113(484): 226–239.

Atkinson, Robert D. 2007. "Expanding the R&E Tax Credit to Drive Innovation, Competitiveness and Prosperity." *Journal of Technology Transfer* 32(6): 617–628.

Audretsch, David B., Albert N. Link, and John T. Scott. 2002. "Public/Private Technology Partnerships: Evaluating SBIR-Supported Research." *Research Policy* 31(1): 145–158.

Audretsch, David B., and A. Roy Thurik. 2001. "What's New about the New Economy? Sources of Growth in the Managed and Entrepreneurial Economies." *Industrial and Corporate Change* 10(1): 267–315.

———. 2004. "A Model of the Entrepreneurial Economy." *International Journal of Entrepreneurship Education* 2(2): 143–166.

Augier, Mie, and David J. Teece. 2007. "Dynamic Capabilities and Multinational Enterprise: Penrosean Insights and Omissions." *Management International Review* 47(2): 175–192.

Baldwin, William L., and John T. Scott. 1987. *Market Structure and Technological Change*. Harwood Fundementals of Pure and Applied Economics Series. Chur, Switzerland: Harwood Academic Publishers.

Baron, James N., Michael T. Hannan, and M. Diane Burton. 1999. "Building the Iron Cage: Determinants of Managerial Intensity in the Early Years of Organizations." *American Sociological Review* 64(4): 527–547.

Baumol, William J. 1990. "Entrepreneurship: Productive, Unproductive, and Destructive." *Journal of Business Venturing* 11(1): 3–22.

Baumol, William J., Robert E. Litan, and Carl J. Schramm. 2007. "Sustaining Entrepreneurial Capitalism." *Capitalism and Society* 2(2): 1–36.

Birch, David L. 1979. *The Job Generation Process*. Cambridge, MA: MIT Program on Neighborhood and Regional Change.

———. 1981. "Who Creates Jobs?" *Public Interest* 65(Fall): 3–14.

———. 1987. *Job Creation in America: How Our Smallest Companies Put the Most People to Work*. New York: Free Press.

Breitzman, Anthony, and Diana Hicks. 2008. *An Analysis of Small Business Patents by Industry and Firm Size*. Washington, DC: Small Business Administration Office of Advocacy.

Carlsson, Bo. 1992. "The Rise of Small Business: Causes and Consequences." In *Singular Europe: Economy and Polity of the European Community after 1992,* William James Adams, ed. Ann Arbor, MI: University of Michigan Press, pp. 145–169.

Clinton, William J., and Albert Gore Jr. 1994. *Science in the National Interest*. Washington, DC: Office of Science and Technology Policy.

Committee on Prospering in the Global Economy of the 21st Century. 2007. *Rising above the Gathering Storm: Energizing and Employing America for a Brighter Economic Future*. Washington, DC: National Academies Press.

Council of Economic Advisers. 1994. *Economic Report of the President*. Washington, DC: U.S. Government Printing Office.

———. 2000. *Economic Report of the President*. Washington, DC: U.S. Government Printing Office.

Davis, Steven J., John Haltiwanger, and Scott Schuh. 1996. "Small Business and Job Creation: Dissecting the Myth and Reassessing the Facts." *Small Business Economics* 8(4): 297–315.

Evans, David S. 1987. "The Relationship between Firm Growth, Size, and Age: Estimates for 100 Manufacturing Industries." *Journal of Industrial Economics* 35(4): 567–581.

Executive Office of the President. 1990. *U.S. Technology Policy*. Washington, DC: Office of Science and Technology Policy.

———. 1997. *Science and Technology: Shaping the Twenty-First Century*. A Report to the Congress. Washington, DC: Office of Science and Technology Policy.

Friedman, Benjamin M. 2007. "Comment on 'Sustaining Entrepreneurial Capitalism' (by William J. Baumol, Robert E. Litan, and Carl J. Schramm)." *Capitalism and Society* 2(2): 1–5.

Greene, William H. 2003. *Econometric Analysis*. 5th ed. Upper Saddle River, NJ: Prentice Hall.

Hall, Bronwyn H., Jacques Mairesse, and Pierre Mohnen. 2010. "Measuring the Returns to R&D." In *Handbook of the Economics of Innovation*, Vol. 2, Bronwyn H. Hall and Nathan Rosenberg, eds. Amsterdam: Elsevier, pp. 1034–1076.

Hébert, Robert F., and Albert N. Link. 1988. 2d ed. *The Entrepreneur: Mainstream Views and Radical Critiques*. New York: Praeger.

———. 2009. *A History of Entrepreneurship*. Routledge Studies in the History of Economics Series. London: Routledge.

Jaffe, Adam B. 1998. "The Importance of 'Spillovers' in the Policy Mission of the Advanced Technology Program." *Journal of Technology Transfer* 23(2): 11–19.

Jewkes, John, David Sawers, and Richard Stillerman. 1958. *The Sources of Invention*. London: Macmillan and Co.

Jovanovic, Boyan. 1982. "Selection and the Evolution of Industry." *Econometrica* 50(3): 649–670.

Kahin, Brian, and Christopher T. Hill. 2010. "United States: The Need for Continuity." *Issues in Science and Technology* 26(3): 51–60.

Kendrick, John W., and Elliot S. Grossman. 1980. *Productivity in the United States: Trends and Cycles*. Baltimore: Johns Hopkins University Press.

Kilby, Peter. 1971. "Hunting the Heffalump." In *Entrepreneurship and Economic Change*, Peter Kilby, ed. New York: Free Press, pp. 1–42.

Klein, Burton H. 1979. "The Slowdown in Productivity Advances: A Dynamic Explanation." In *Technological Innovation for a Dynamic Economy,* Christopher T. Hill and James M. Utterback, eds. New York: Pergamon Press, pp. 66–117.

Knight, Frank H. 1921. *Risk, Uncertainty, and Profit*. Boston: Hart, Schaffner, and Marx.

Kohn, Meir, and John T. Scott. 1982. "Scale Economies in Research and Development: The Schumpeterian Hypothesis." *Journal of Industrial Economics* 30(3): 239–249.

Lerner, Josh. 1999. "The Government as Venture Capitalist: The Long-Run Impact of the SBIR Program." *Journal of Business* 72(3): 285–318.

———. 2010. "The Future of Public Efforts to Boost Entrepreneurship and Venture Capital." *Small Business Economics* 35(3): 255–264.

Lerner, Josh, and Colin Kegler. 2000. "Evaluating the Small Business Innovation Research Program: A Literature Review." In *The Small Business Innovation Research Program: An Assessment of the Department of Defense Fast Track Initiative*, Charles W. Wessner, ed. Washington, DC: National Academy Press, pp. 307–324.

Link, Albert N. 1987. *Technological Change and Productivity Growth*. Harwood Fundamentals of Pure and Applied Economics Series, No. 13. Chur, Switzerland: Harwood Academic Publishers.

———. 1999. "Public/Private Partnerships in the United States." *Industry and Innovation* 6(2): 191–217.

———. 2006. *Public/Private Partnerships: Innovation Strategies and Policy Alternatives*. New York: Springer.

Link, Albert N., and John T. Scott. 2000. "Estimates of the Social Returns to Small Business Innovation Research Projects." In *The Small Business Innovation Research Program: An Assessment of the Department of Defense Fast Track Initiative*, Charles W. Wessner, ed. Washington, DC: National Academy Press, pp. 275–290.

———. 2001. "Public/Private Partnerships: Stimulating Competition in a Dynamic Market." *International Journal of Industrial Organization* 19(5): 763–794.

———. 2003. "U.S. Science Parks: The Diffusion of an Innovation and Its Effects on the Academic Missions of Universities." *International Journal of Industrial Organization* 21(9): 1323–1356.

———. 2006. "U.S. University Research Parks." *Journal of Productivity Analysis* 25(1–2): 43–55.

———. 2009. "Private Investor Participation and Commercialization Rates for Government-Sponsored Research and Development: Would a Prediction Market Improve the Performance of the SBIR Programme?" *Economica* 76(302): 264–281.

———. 2010. "Government as Entrepreneur: Evaluating the Commercialization Success of SBIR Projects." *Research Policy* 39(5): 589–601.

———. 2011. *Public Goods, Public Gains: Calculating the Social Benefits of Public R&D*. New York: Oxford University Press.

Link, Albert N., and Donald S. Siegel. 2003. *Technological Change and Economic Performance*. London: Routledge.

———. 2007. *Innovation, Entrepreneurship, and Technological Change*. Oxford: Oxford University Press.

Link, Albert N., and Gregory Tassey. 1987. *Strategies for Technology-Based Competition: Meeting the New Global Challenge*. Lexington, MA: Lexington Books.

Martin, Stephen, and John T. Scott. 2000. "The Nature of Innovation Market Failure and the Design of Public Support for Private Innovation." *Research Policy* 29(4–5): 437–447.

National Association of Seed and Venture Funds. 2006. *Seed and Venture Capital: State Experiences and Options*. Chicago, IL: National Association of Seed and Venture Funds.

National Economic Council. 2009. *A Strategy for American Innovation: Driving towards Sustainable Growth and Quality Jobs*. Washington, DC: Executive Office of the President.

Nelson, Richard R. 1982. "Government Stimulus of Technological Progress: Lessons from American History." In *Government and Technical Progress:*

A Cross-Industry Analysis, Richard R. Nelson, ed. Technology Policy and Economic Growth Series. New York: Pergamon Press, pp. 451–482.

Office of Management and Budget. 1992. *Circular No. A-94 Revised: Guidelines and Discount Rates for Benefit-Cost Analysis of Federal Programs*. Washington, DC: Executive Office of the President.

———. 1996. *Economic Analysis of Federal Regulations under Executive Order 12866*. Washington, DC: Executive Office of the President.

———. 2010. *Analytical Perspectives: Budget of the U.S. Government, Fiscal Year 2011*. Washington, DC: Government Printing Office.

Orszag, Peter R. 2009. "OMB Memorandum: Increased Emphasis on Program Evaluations." Washington, DC: Executive Office of the President.

Public Broadcasting Service (PBS). 2002. *Rediscovering George Washington: Other Documents—First Annual Message to Congress, January 8, 1790*. Arlington, VA: Public Broadcasting Service. http://www.pbs.org/georgewashington/collection/other_1790jan8.html (accessed July 29, 2011).

Rausch, Lawrence M. 2010. "Indicators of U.S. Small Business's Role in R&D." *InfoBrief*, NSF 10-304: 1–6.

Richardson, George B. 1972. "The Organisation of Industry." *Economic Journal* 82(327): 883–896.

Sargent, John F. Jr. 2010. *America COMPETES Act and the FY2010 Budget*. CRS Report for Congress. Washington, DC: Congressional Research Service.

Schacht, Wendy H. 2011. *The Small Business Innovation Research (SBIR) Program: Reauthorization Efforts*. CRS Report for Congress. Washington, DC: Congressional Research Service.

Schultz, Theodore W. 1980. "Investment in Entrepreneurial Ability." *Scandinavian Journal of Economics* 82(4): 437–448.

Schumpeter, Joseph A. 1934. *The Theory of Economic Development*. Cambridge, MA: Harvard University Press.

———. 1950. *Capitalism, Socialism and Democracy*. New York: Harper and Brothers.

Scott, John T. 1984. "Firm versus Industry Variability in R&D Intensity." In *R&D, Patents, and Productivity*, Zvi Griliches, ed. Chicago: University of Chicago Press, pp. 233–245.

———. 2008. "The National Cooperative Research and Production Act." In *Issues in Competition Law and Policy*, Vol. 2, Wayne Dale Collins, ed. Chicago: American Bar Association, pp. 1297–1317.

Scott, John T., and George Pascoe. 1986. "Beyond Firm and Industry Effects on Profitability in Imperfect Markets." *Review of Economics and Statistics* 68(2): 284–292.

Solow, Robert M. 1957. "Technical Change and the Aggregate Production Function." *Review of Economics and Statistics* 39(3): 312–320.

StataCorp. 2007. *Stata Statistical Software: Release 10*. College Station, TX: StataCorp LP.

Stevens, Ashley J. 2004. "The Enactment of Bayh-Dole." *Journal of Technology Transfer* 29(1): 93–99.

Stinchcombe, Arthur L. 1965. "Social Structure and Organizations." In *Handbook of Organizations*, James G. March, ed. Chicago: Rand McNally and Company, pp. 142–193.

Stine, Deborah D. 2009. *America COMPETES Act: Programs, Funding, and Selected Issues*. CRS Report for Congress. Washington, DC: Congressional Research Service.

Sutton, John. 1997. "Gibrat's Legacy." *Journal of Economic Literature* 35(1): 40–59.

Tassey, Gregory. 1992. *Technology Infrastructure and Competitive Position*. Norwell, MA: Kluwer Academic Publishers.

———. 1997. *The Economics of R&D Policy*. Westport, CT: Quorum Books.

———. 2003. *Methods for Assessing the Economic Impacts of Government R&D*. NIST Planning Report No. 03-1. Gaithersburg, MD: National Institute of Standards and Technology.

———. 2005. "Underinvestment in Public Good Technologies." *Journal of Technology Transfer* 30(1–2): 89–113.

———. 2007. "Tax Incentives for Innovation: Time to Restructure the R&E Tax Credit." *Journal of Technology Transfer* 32(6): 605–615.

Thurik, A. Roy. 2009. "Entreprenomics: Entrepreneurship, Economic Growth, and Policy." In *Entrepreneurship, Growth, and Public Policy,* Zolton J. Acs, David B. Audretsch, and Robert J. Strom, eds. Cambridge: Cambridge University Press, pp. 219–249.

Tibbetts, Roland. 1999. "The Small Business Innovation Research Program and NSF SBIR Commercialization Results." In *SBIR—The Small Business Innovation Research Program: Challenges and Opportunities*, Charles W. Wessner, ed. Washington, DC: National Academy Press, pp. 129–167.

U.S. Census Bureau. 2010. *Census Regions and Divisions of the United States*. Washington, DC: U.S. Census Bureau. http://www.census.gov/geo/www/us_regdiv.pdf (accessed June 28, 2011).

U.S. Senate. 2008. *SBIR/STTR Reauthorization Act of 2008*. Senate Report 110-447. Washington, DC: U.S. Government Printing Office.

Wallsten, Scott J. 2000. "The Effects of Government-Industry R&D Programs on Private R&D: The Case of the Small Business Innovation Research Program." *RAND Journal of Economics* 31(1): 82–100.

Wessner, Charles W., ed. 2000. *The Small Business Innovation Research Pro-*

gram: An Assessment of the Department of Defense Fast Track Initiative. Washington, DC: National Academy Press.

———. 2004. *SBIR Program Diversity and Assessment Challenges: Report of a Symposium*. Washington, DC: National Academies Press.

———. 2008. *An Assessment of the SBIR Program*. Washington, DC: National Academies Press.

Zeira, Joseph. 1987. "Investment as a Process of Search." *Journal of Political Economy* 95(1): 204–210.

Authors

Albert N. Link is a professor of economics at the University of North Carolina at Greensboro (UNCG). He received a BS in mathematics from the University of Richmond in 1971 and a PhD in economics from Tulane University in 1976. After earning the PhD, he joined the economics faculty at Auburn University, where he remained until 1982, when he joined the economics faculty at UNCG. While at UNCG, he has served as head of the department of economics as well as director of the MBA program.

Link's research focuses on innovation policy, academic entrepreneurship, and the economics of research and development. He has served as editor-in-chief of the *Journal of Technology Transfer* since 1996. Among his most recent books are *Government as Entrepreneur* (Oxford University Press, 2009), *Cyber Security: Economic Strategies and Public Policy Alternatives* (Edward Elgar, 2008), and *Entrepreneurship, Innovation, and Technological Change* (Oxford University Press, 2007). His other research has appeared in such journals as the *American Economic Review*, the *Journal of Political Economy*, the *Review of Economics and Statistics*, *Economica*, and *Research Policy*.

Much of Link's research has been supported by funding organizations such as the National Science Foundation, the Organisation for Economic Co-operation and Development (OECD), the World Bank, and the science and technology ministries in several developed nations. Currently, Link is serving as the vice-chairperson of the Team of Specialists on Innovation and Competitiveness Policies for the United Nations Economic Commission for Europe (UNECE).

John T. Scott is a professor of economics at Dartmouth College. He earned a PhD in economics from Harvard University and an AB in economics and English from the University of North Carolina at Chapel Hill. His research is in the areas of industrial organization and the economics of technological change. He has served as president of the Industrial Organization Society and on the editorial boards of the *International Journal of Industrial Organization*, the *Review of Industrial Organization*, and the *Journal of Industrial Economics*. He has also served as an economist for the Board of Governors of the Federal Reserve System and for the Federal Trade Commission. Scott's books include *Market Structure and Technological Change* (Harwood Academic, 1987), *Purposive Diversification and Economic Performance* (Cambridge University Press, 1993, 2005), and *Environmental Research and Development: U.S. Industrial Research, the Clean Air Act, and Environmental Damage* (Edward Elgar, 2003). His published research in academic journals has spanned four decades, and his research about the economics of technological change has been supported by the National Institute of Standards and Technology, the National Science Foundation, the National Academy of Sciences, the World Bank, the United Nations Development Programme, and the Organisation for Economic Co-operation and Development.

Index

The italic letters *f, n,* and *t* following a page number indicate that the subject information of the heading is within a figure, note, or table, respectively, on that page. Double italics indicate multiple but consecutive elements.

About the Institute

The W.E. Upjohn Institute for Employment Research is a nonprofit research organization devoted to finding and promoting solutions to employment-related problems at the national, state, and local levels. It is an activity of the W.E. Upjohn Unemployment Trustee Corporation, which was established in 1932 to administer a fund set aside by Dr. W.E. Upjohn, founder of The Upjohn Company, to seek ways to counteract the loss of employment income during economic downturns.

The Institute is funded largely by income from the W.E. Upjohn Unemployment Trust, supplemented by outside grants, contracts, and sales of publications. Activities of the Institute comprise the following elements: 1) a research program conducted by a resident staff of professional social scientists; 2) a competitive grant program, which expands and complements the internal research program by providing financial support to researchers outside the Institute; 3) a publications program, which provides the major vehicle for disseminating the research of staff and grantees, as well as other selected works in the field; and 4) an Employment Management Services division, which manages most of the publicly funded employment and training programs in the local area.

The broad objectives of the Institute's research, grant, and publication programs are to 1) promote scholarship and experimentation on issues of public and private employment and unemployment policy, and 2) make knowledge and scholarship relevant and useful to policymakers in their pursuit of solutions to employment and unemployment problems.

Current areas of concentration for these programs include causes, consequences, and measures to alleviate unemployment; social insurance and income maintenance programs; compensation; workforce quality; work arrangements; family labor issues; labor-management relations; and regional economic development and local labor markets.

About the Institute